CONCEPTS IN CHEMISTRY (CHM 2450) WORKBOOK FOR PRE-SERVICE ELEMENTARY AND MIDDLE SCHOOL TEACHERS

DR. BASIL MUGAGA NAAH

WRIGHT STATE UNIVERSITY

MASTERCHEM PUBLISHING COMPANY

3007 Eastham Street, Springfield, OH 45503

TABLE OF CONTENTS

PREFACE

The best way to learn chemistry is to do chemistry experiments, take part in group discussions, and ask questions in class and outside of class about natural phenomena.

The chemistry concepts discussed in this workbook are aligned with the Ohio Science Standards and the Next Generation Science Standards.

This chemistry workbook will help you study basic concepts in chemistry through group discussions and hands-on experiments.

The concept maps you generate in the workbook will help you discover relationships between chemistry concepts. These relationships will help you improve your understanding of chemistry.

The chemistry misconceptions research paper will help you identify chemistry misconceptions common among Elementary and Middle school students.

The teaching demonstration will help you learn the best practices on how to address these misconceptions. And the hands-on activities will help make abstract chemistry concepts concrete so that you can make sense of these concepts as you learn them.

ACKNOWLEDGMENTS

I want to thank Brandyn Kim Jackson who was a former student of this course for his helpful suggestions.

TO PRE-SERVICE TEACHERS

Welcome to concepts in chemistry (CHM 2450) for pre-service elementary and middle school teachers. This workbook contains experiments, chemistry concepts, concept maps, teaching demonstration, and misconceptions research assignment. Under each topic, you will find learning objectives. These learning objectives are to help you structure your learning. Carefully read these learning objectives before and after class as you study the material in this course.

The experiments in the workbook are guided inquiry, and they are based on everyday phenomena we observe around us. You are to work together in small groups to do these experiments and answer the corresponding postlab questions. Once you understand the concepts in these experiments, you can apply them to design your own experiments for your future classroom.

NEXT GENERATION SCIENCE STANDARDS (NGSS)

LEARNING OBJECTIVES

After this lesson, you should be able to explain:

- crosscutting concepts
- disciplinary core areas
- science and engineering practices

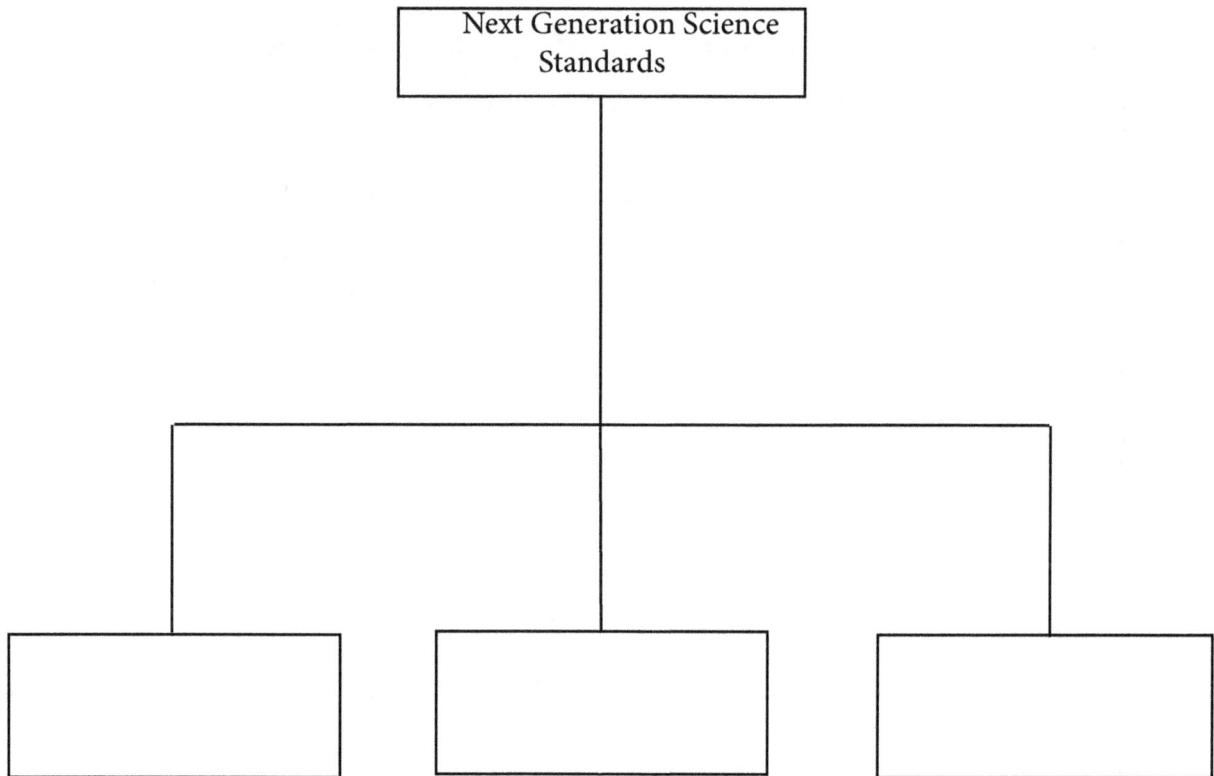

```
┌─────────────────────────┐
│ Next Generation Science │
│        Standards        │
└─────────────────────────┘
          │
  ┌───────┼───────┐
  │       │       │
┌────┐  ┌────┐  ┌────┐
│    │  │    │  │    │
└────┘  └────┘  └────┘
```

1.   What are disciplinary core ideas? Give one example in chemistry.

2.   What are crosscutting concepts? List and explain all seven.

3.   What are science and engineering practices? Give examples of these practices.

THE THREE LEVELS OF REPRESENTATION IN CHEMISTRY

LEARNING OBJECTIVES

After this lesson, you should be able to describe:

- macroscopic level
- symbolic level
- microscopic level

ACTIVITY 1.0: DESCRIBE THE ROCK ACCORDING TO THE THREE LEVELS

1.   Macroscopic level (What can you perceive with your senses?).

2.   Microscopic level (What can you not perceive with your senses but it's there?).

3.   Symbolic level (What symbols can you use to describe the rock?)

HOW TO APPLY THE THREE LEVELS OF REPRESENTATION TO TEACH CHEMISTRY

1.   How will you apply the three levels of representation to teach the three states of matter?

HOW TO APPLY SCIENTIFIC PRACTICES TO EXPLORE
THE LAW OF CONSERVATION OF MASS

LEARNING OBJECTIVES

After this lesson, you should be able to:

- explain the law of conservation of mass
- explain hypothesis, law, theory, and experiment

REACTION

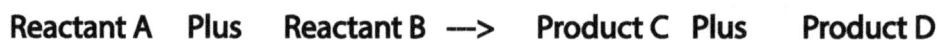

Reactant A   Plus   Reactant B --->   Product C Plus   Product D

Describe at:

Macroscopic :

Symbolic:

Microscopic:

Question

1.   When reactant A reacts with reactant B, will the mass of the product be

    a. Less than the combined masses of reactant A and reactant B

    b. Equal to the combined masses of reactant A and reactant B

    c. Greater than the combined masses of reactant A and reactant B

2.   Explain why (hypothesis) you picked a particular answer?

3.   How will you test your hypothesis?

UNITS AND MEASUREMENTS

LEARNING OBJECTIVES

After this lesson, you should be able to:

- define exact and inexact numbers
- explain random and systematic errors
- explain precision and accuracy
- set up conversion factors
- convert from one form of unit to another

THE TWO KINDS OF NUMBERS

In measurement, there are two kinds of numbers. These numbers are: exact and inexact

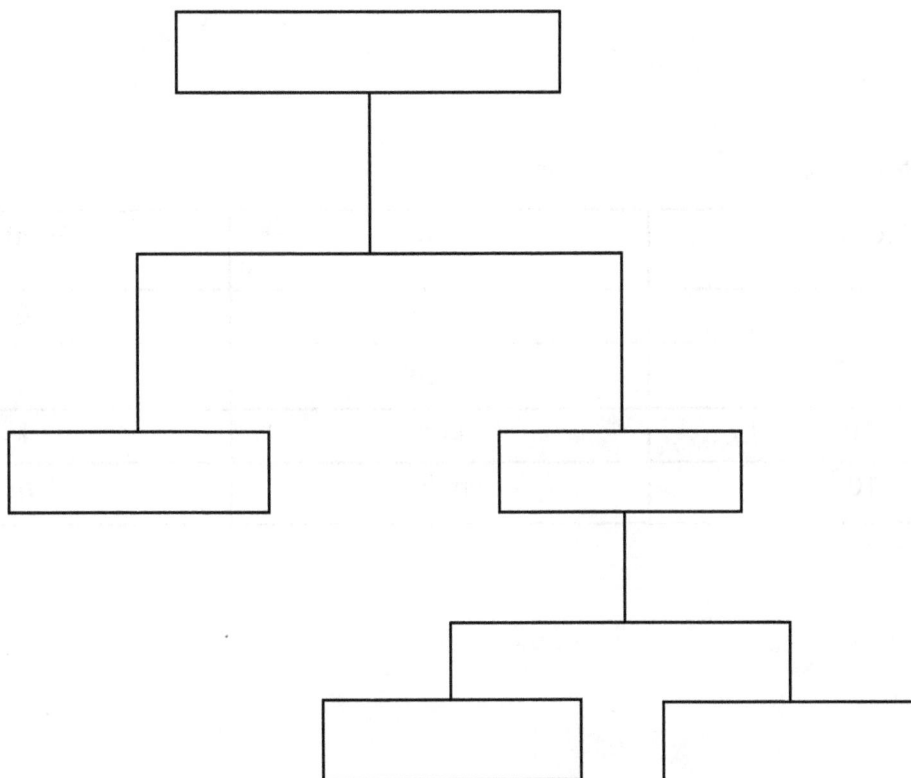

Table 1: SI units

Quantity measured	Symbol	Unit name	Symbol
length	l	meter	m
mass	m	kilogram	kg
time	t	Second	s
temperature	T	Kelvin	K
amount of substance	n	mole	mol

Note: SI stands for International System of Units

Table 2: SI Prefixes

Multiple	Prefix	Symbol
10^9	giga	G
10^6	mega	M
10^3	kilo	k
10^{-3}	milli	m

QUESTIONS ON MEASUREMENT

1.   What are systematic errors?

2.   How can systematic errors be addressed?

3.   What are random errors?

4.   How can random errors be addressed

PRECISION AND ACCURACY

1. What is precision?

2. What is accuracy?

Use the diagrams below to answer questions 3 to 6

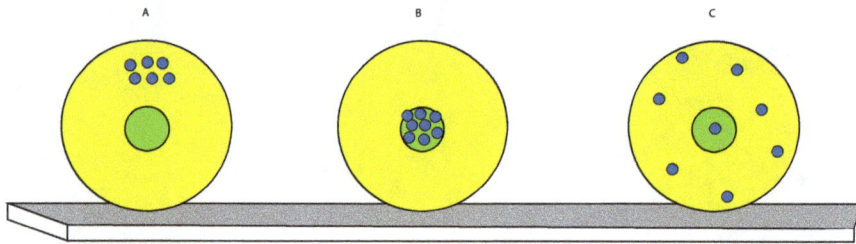

3. Which diagrams illustrate precision? Explain

4. Which diagram illustrates accuracy? Explain

5. Which diagram illustrates neither precision nor accuracy? Explain

6. Which diagram illustrates precision but not accuracy? Explain

7.   What are derived units?

8.   What are alternative units?

USE THE TABLE BELOW TO ANSWER THE QUESTIONS THAT FOLLOW

Number of Trials	Student A	Student B	Student C
1	2.335 g	2.357 g	2.369 g
2	2.331g	2.375 g	2.373 g
3	2.333 g	2.338 g	2.371 g

9.   Calculate the average value for :

         Student A

         Student B

         Student C

10.   If the actual value is 2.36 g, how will you evaluate each student in terms of

         accuracy

         precision

HOW TO CONVERT FROM ONE UNIT TO ANOTHER

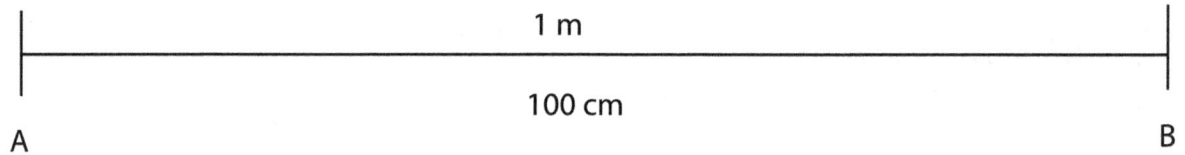

1 m
_____
100 cm

A   B

Assume that the distance from A to B when measured in meters is equal to 1 m. Assume that this same distance when measured in cm is equal to 100 cm. Therefore, since the distance between A and B hasn't changed, we can write that 100 cm = 1 m. From this relationship, when we divide both sides of the equal sign by 1 m, we will generate the ratio $\dfrac{100\ cm}{1 cm}$.

And when we divide both sides by 100 cm, we will generate the ratio $\dfrac{1\ m}{100\ cm}$

These ratios are called conversion factors. We use them to convert from cm to m or from m to cm.

Conversion factors

1. Write first conversion factor here:

2. Write second conversion factor here:

Example 1

How many cm are in 2 m ?

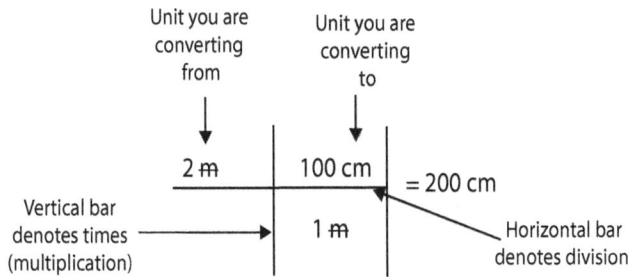

Unit you are converting from ↓ 2 m

Unit you are converting to ↓ 100 cm

$= 200$ cm

Vertical bar denotes times (multiplication) →

1 m

Horizontal bar denotes division

We can rewrite the above expression as:

$$2\ m \times \frac{100\ cm}{1\ m} = 200\ cm$$

This approach helps you keep track of units in a clean way as you convert

Example 2

How many minutes are in 1 hour?

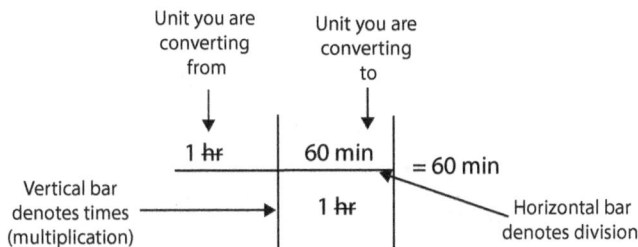

Unit you are converting from ↓ 1 hr

Unit you are converting to ↓ 60 min

$= 60$ min

Vertical bar denotes times (multiplication) →

1 hr

Horizontal bar denotes division

We can rewrite the above expression as:

$$1\ hr \times \frac{60\ min}{1\ hr} = 60\ min$$

This approach helps you keep track of units in a clean way as you convert

Example 3

How many seconds are in 1 hour?

CONVERT FROM ONE UNIT TO ANOTHER

1.

    a. 2500 cm to m (100 cm = 1m)

    b. 3455 g to kg (1000g = 1kg)

    c. 40 cm to mm (10 mm = 1cm)

    d. 245 mL to L (1000 mL = 1 L)

2. Covert 3 **mol** of H_2O to **molecules** of H_2O (1mol of H_2O = 6.02 x 10^{23} molecules of H_2O)

3. Convert 5 **mol** of O_2 to **liters** (L) of O_2 (1 mol = 22.4 L)

ACTIVITY 2: WEIGHING BY THE DIFFERENCE METHOD

OBJECTIVE

Students will use the difference method to determine the mass of an object

PROCEDURE

Obtain a beaker and select one metal block from the instructor's desk. Go to a balance, weigh your empty beaker and record its mass in Table 1 below. Next, put your metal object into the beaker, weigh again and record its mass. Finally, use the two values from your weighing to determine the mass of your object. Keep in mind that you can use this same approach to determine the mass of liquid.

Table 1: Finding the mass of an object by the difference method

Object	Mass, g
Mass of empty beaker	
Mass of empty beak plus mass of object	
Mass of object	

POST-ACTIVITY QUESTIONS

1.   Explain the difference method

2.   What's the difference between mass and weight?

MEASURING THE AMOUNT OF SPACE AN OBJECT TAKES UP (VOLUME)

In measurement, volume has two meanings: first, it can refer to the amount of space an object occupies and second the amount of something an object can hold (capacity).

OBJECTIVE

Students will use a mathematical formula and the water-displacement method to determine the volume of an object. They will then compare these values to see if they are equal.

Part A: Using a mathematical formula to determine the volume of an object

First use the caliper or ruler to measure the height, length and width of the metal block. Multiply these numbers together to determine the volume of the metal block.

Part B: Using the water-displacement method to determine the volume of an object

Put about 30 mL of water in a graduated cylinder; record this value as your initial volume of liquid in the cylinder. Drop your metal block into the graduated cylinder, and record the new volume of water as your final volume. Now subtract your initial volume from the final volume to get the volume of the object.

POST-ACTIVITY QUESTIONS

1.   Was the volume of the metal block the same for both approaches? Explain.

2.   Can you use water-displacement to determine the volume of an object that floats in water? Explain.

DENSITY

LEARNING OBJECTIVES

After this lesson, you should be able to:

- define density
- calculate the density of a substance
- draw particle models to illustrate the density of an object

What is Density?

How to determine the volume of rectangular block (regular object).

How to determine the volume of rock (irregular object)

1.   A graduated cylinder contains 30.0 mL of water. A student drops a rock of mass 15.5 g

into the cylinder, and the water level rises to 45.0 cm³. Calculate the density of the rock.

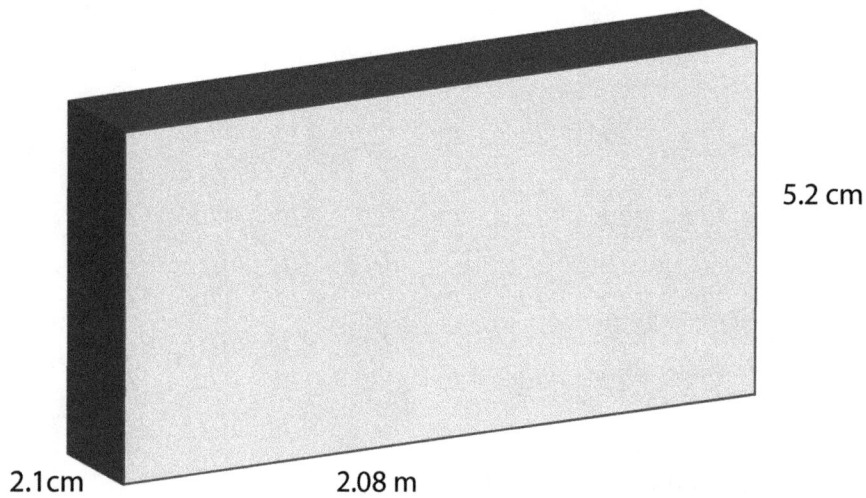

5.2 cm   **Figure 1**

2.1cm   2.08 m

2.   If the mass of the rectangular block in figure 1 is 25 kg, calculate its density.

EXPERIMENT 1: WILL DIET OR REGULAR SODA SINK OR FLOAT IN WATER

OBJECTIVE

In this experiment, you will explore why regular or diet soda sinks or floats in water.

PREDICTION

Will diet or regular soda sink or float in water? Explain.

MATERIALS

Water, a 1000 mL beaker, cans of diet and regular soda.

PROCEDURE

Develop a procedure to determine whether diet or regular soda will float or sink in water.

POST-LAB QUESTIONS

1. Why is it that both cans (regular and diet soda) have the same volume of soda, but one can floats while the other sinks? Include evidence to support your answer.

2. Are the ingredients in diet similar to those in regular soda? Explain.

3. Is the density of diet or regular soda greater than that of water? Explain.

BLANK WORKSHEET

EXPERIMENT 2: DETERMINE THE DENSITY OF SOLUTION AND PENNIES

PREDICTIONS

Will the density of diet be greater than the density of regular? Explain.

Will the density of a penny be greater than the density of regular soda? Explain.

PART A: DETERMINE THE DENISTY OF SOLUTION

Develop a procedure to determine the density of regular and diet coke solutions.

POST-LAB QUESTIONS

Compare the results of this experiment to the results of the experiment : will diet or regular soda float or sink in water? Do you see anything in this experiment that confirms or refutes the results of will it sink or float? Explain. What are some sources of error in this experiment? Explain.

PART B: DETERMINE THE DENISTY OF 20 PENNIES

Develop a procedure to determine the density of 20 pennies.

If the actual density (pennies) is 6.85 g/mL, calculate the percentage error.

POST-LAB QUESTIONS

How well did you do the experiment? What are some sources of error and why?

BLANK WORKSHEET

HOW TO IDENTIFY MATTER

LEARNING OBJECTIVES

After this lesson, you should be able to define and cite examples of:

- physical properties
- chemical properties

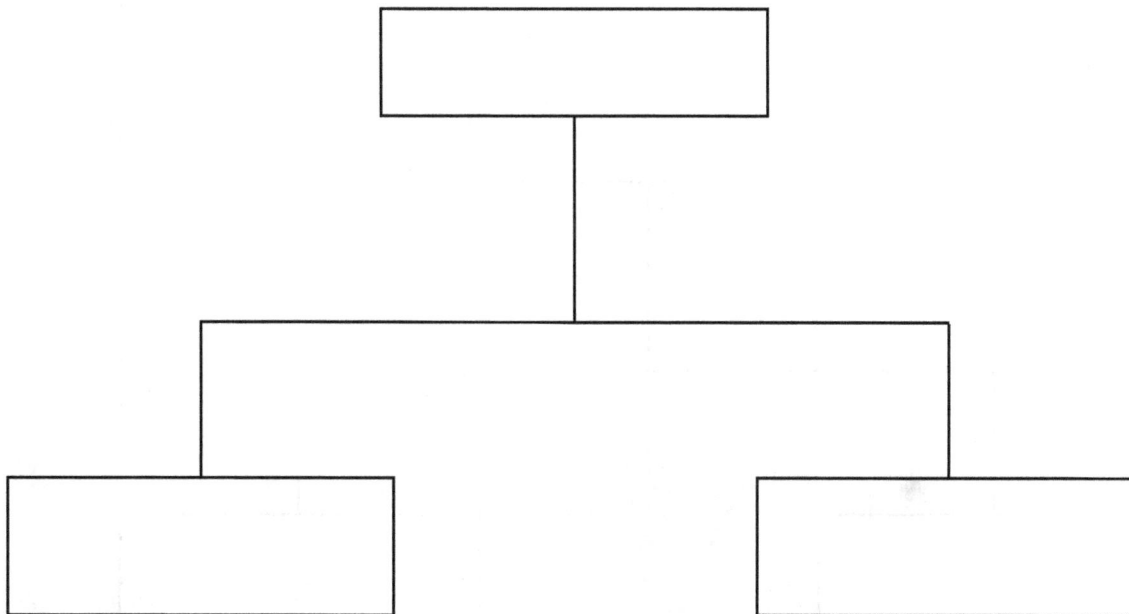

1. What is matter?

2. What are physical properties? Cite an example.

3. What are chemical properties? Cite an example.

TYPES OF CHANGE MATTER CAN UNDERGO

LEARNING OBJECTIVES

After this lesson, you should be able to:

- define and cite examples of physical change
- define and cite examples of chemical change
- Identify physical or chemical change from molecular diagrams

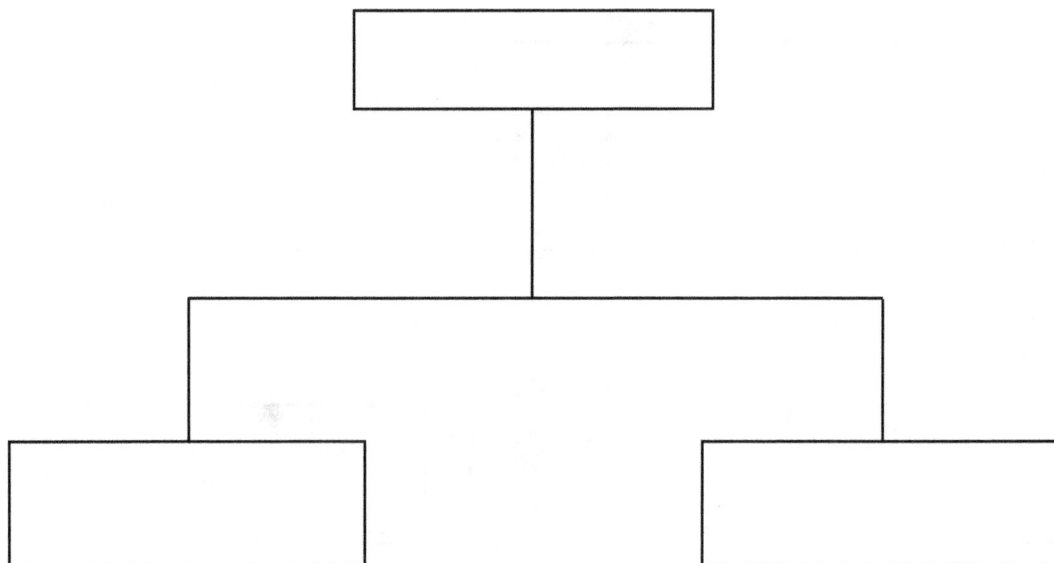

1.   What is a physical change? Cite an example.

2.   What is a chemical change? Cite an example.

PRACTICE PROBLEMS

1. From the molecular diagram below, chemical A can undergo two types of change.

    Identify and explain the type of change for path B and C.

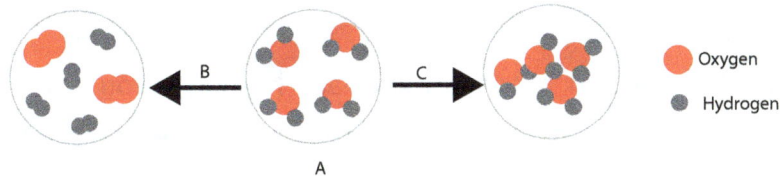

2. From the molecular diagram below, identify the type of change when steam turns into

    ice. Explain the reasoning behind your answer.

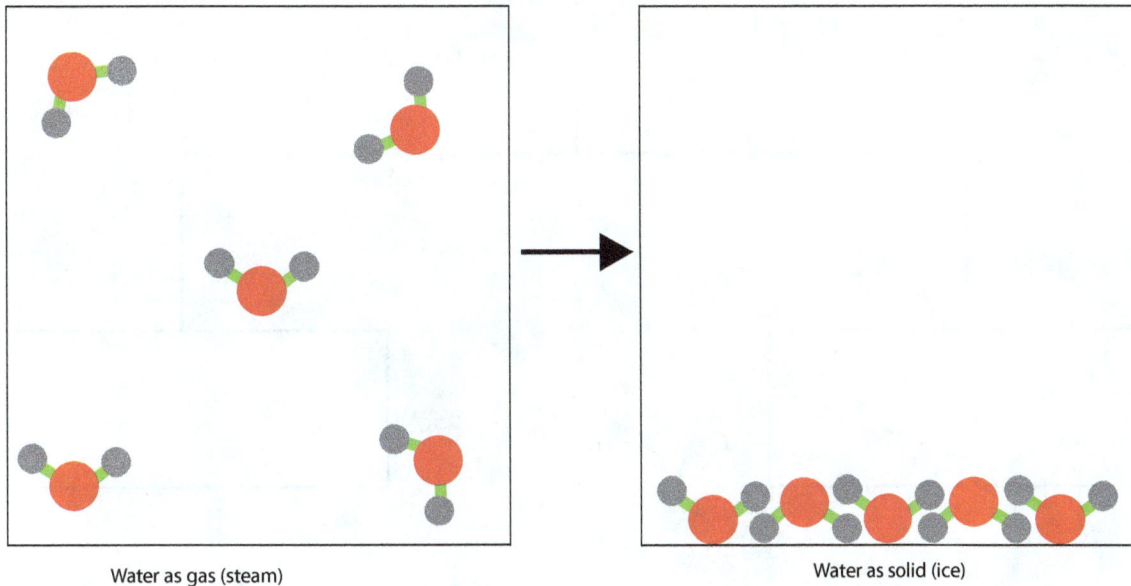

Water as gas (steam)   Water as solid (ice)

LEARNING OBJECTIVES

After this lesson, you should be able to:

- state the three states of matter
- describe the general properties of solids, liquids, and gases
- define pure substance and mixture
- define an element, compound, heterogeneous and homogeneous mixture
- draw particle models to describe pure substance, mixture, element, compound, and molecule
- describe common chemical techniques used to separate mixtures

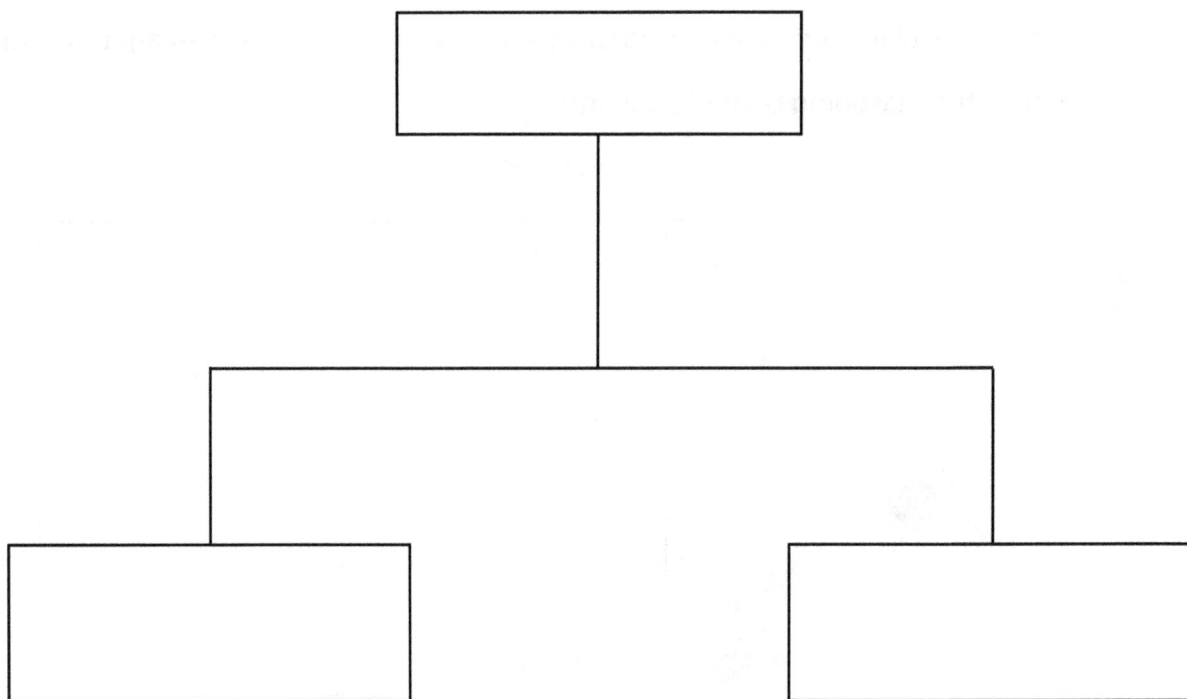

HOW TO CLASSIFY MATTER BY STATE

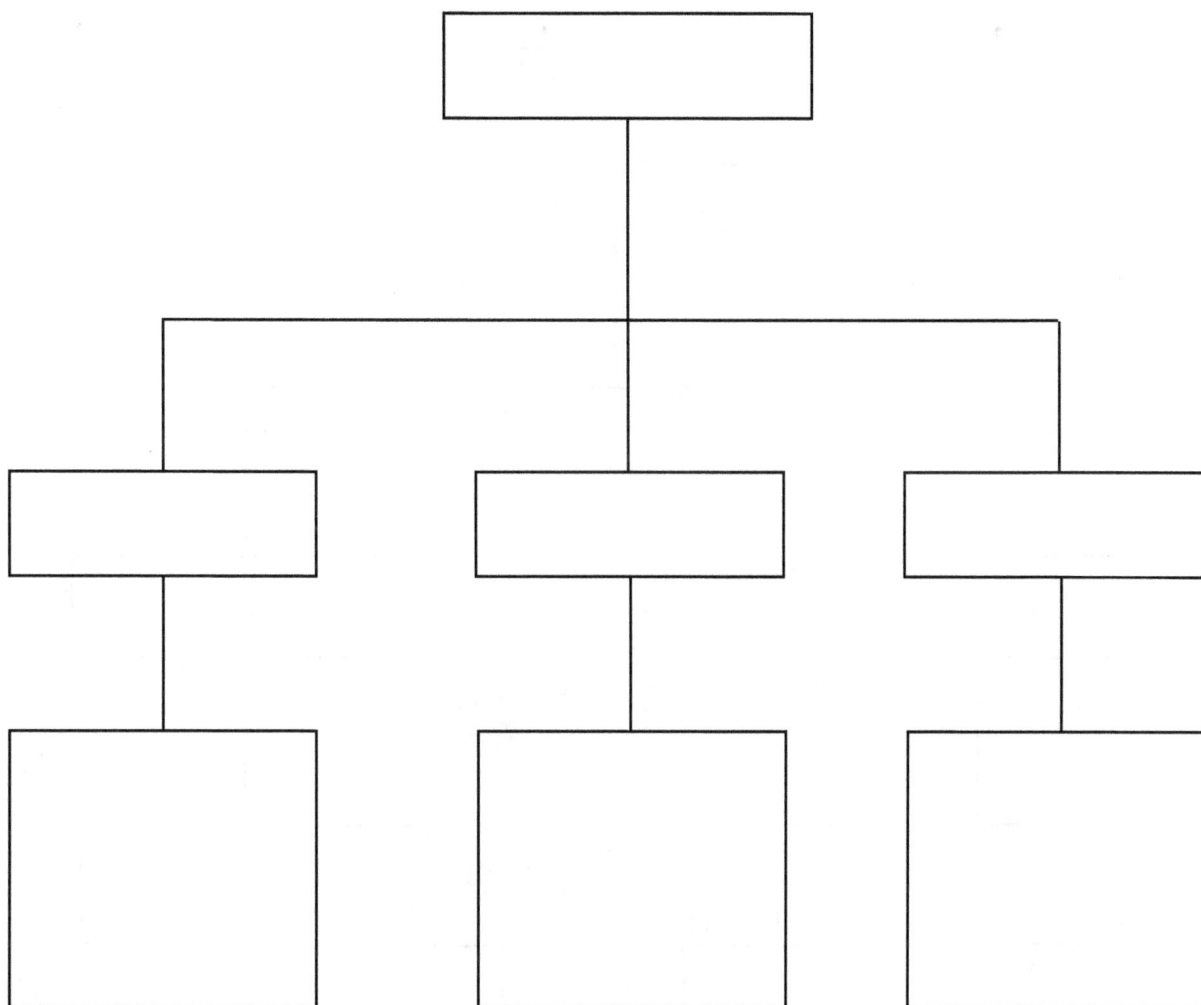

HOW TO CLASSIFY MATTER BY COMPOSITION

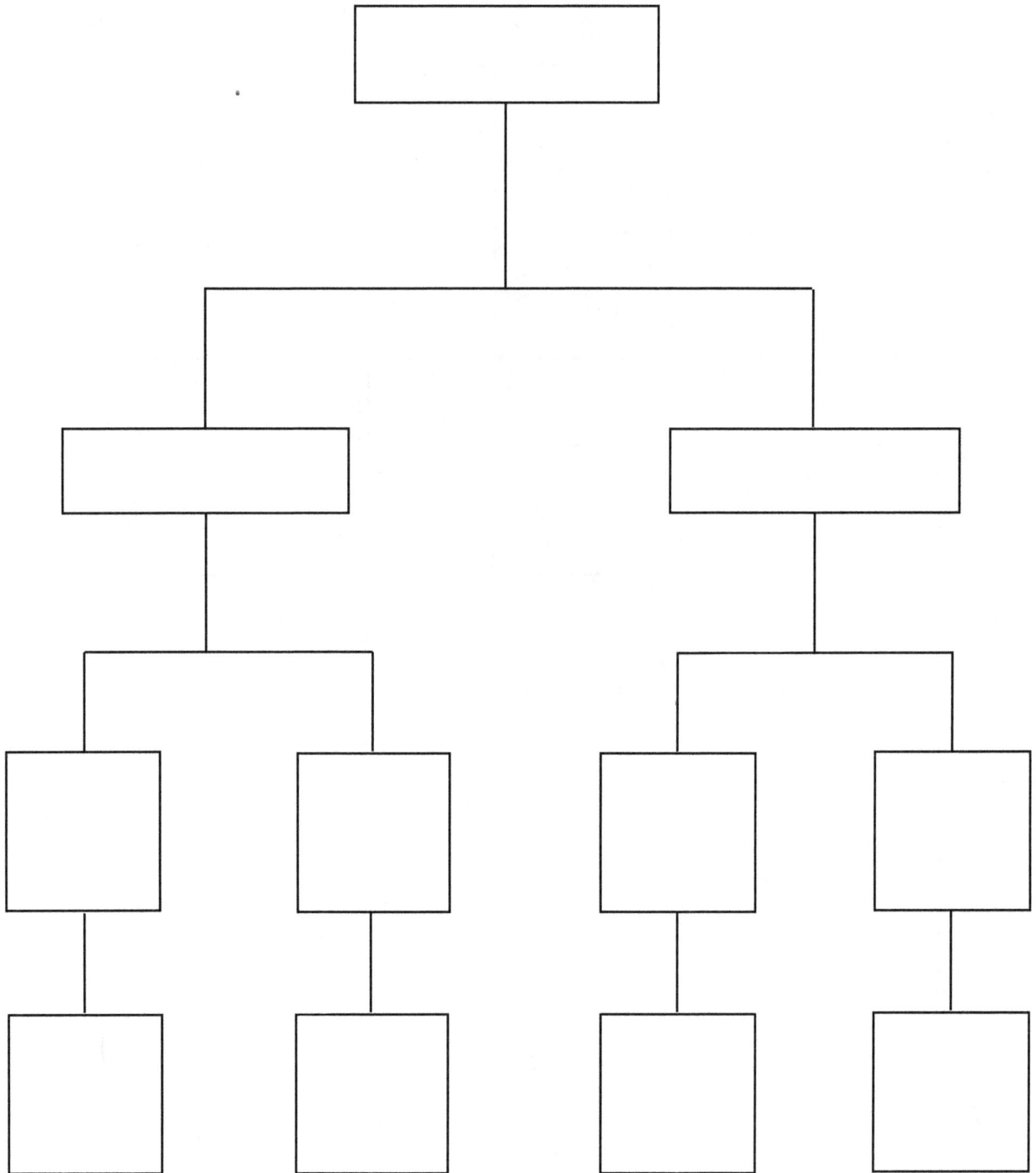

ACTIVITY 2.0: MATCH EACH TERM TO ONE OF THE DEFINITIONS FROM (1) THROUGH (8)

TERMS

Atom; Element; Molecule; Compound; Mixture; Pure substance; Heterogeneous mixture; Homogenous mixture.

DEFINITIONS

1. Consists of the same type of atoms.

2. The smallest piece of an element that has the chemical properties of the element and can take part in a chemical reaction.

3. When at least two different atoms are bonded to form a substance with fixed composition.

4. When chemicals in a mixture are not uniformly distributed throughout the mixture.

5. When chemicals in a mixture are uniformly distributed throughout the mixture.

6. When a substance consists of the same type of chemical with fixed compositon throughout.

7. When two or more atoms are bonded together, and it is the smallest piece of a compound.

8. When two or more chemicals are intermingled with one another.

ACTIVITY 2.0: DRAW PARTICLE MODELS TO DESCRIBE THE COMPOSITION OF MATTER

1. Draw a particle model for an atom in the box:

    Explain why your model depicts an atom.

2. Draw a particle model for an element in the box:

    Explain why your model depicts an element.

3. Draw a particle model for a compound in the box:

    Explain why your model depicts a compound.

4. Draw a particle model for a heterogeneous mixture in the box:

    Explain why your model depicts a heterogeneous mixture.

5. Draw a particle model for a homogenous mixture in the box:

    Explain why your model depicts a homogeneous mixture.

5.   Draw a particle model for a molecule in the box:

Explain why your model depicts a molecule.

5.   Draw a particle model for a pure substance in the box:

Explain why your model depicts a pure substance.

HOW TO SEPARATE A MIXTURE

1.   Describe some everyday chemical techniques you can use to separate a mixture?

ACTIVITY: 3.0: DESCRIBE THE DIFFERENCES BETWEEN THE THREE STATES OF MATTER

Examine the properties of copper metal block, water in beaker and in flask, air in syringe and in balloon. From your observations, answer the questions in the following table.

Property	Solid	Liquid	Gas
Volume			
Has it got fixed volume?			
Shape			
Has it got fixed shape?			
Flow			
Can it easily flow			
Compress			
Can it easily be compressed?			
Molecular level			
How are the molecules arranged? *(Include particle models)*			
At the molecular level, explain why the molecules are arranged that way (remember to consider the strength of kinetic energy and force of attraction between molecules)			

EXPERIMENT 3: EXPLORING THE HEATING CURVE AND PHASE CHANGES OF WATER

Materials: beaker, ice, water, hotplate, test tube, temperature sensor, and graphing calculator.

PREDICTIONS

a) Predict whether energy is transferred to or from ice as it's heated. What're the reasons for your prediction?

b) Predict whether the temperature of ice will increase continuously or will not increase continuously as ice is heated. What're the reasons for your prediction?

c) Draw a graph of temperature versus time as ice is heated.

PROCEDURE

Design a procedure and check with your instructor before carrying out the experiment.

POST-LAB QUESTIONS

1. Draw and label the graph of temperature versus time as ice is converted to steam.

2. Describe the physical changes and the temperature at which these changes occur.

3. If energy is transferred from hotplate to ice, explain why the temperature remains constant at the freezing and boiling points.

4. Draw a particle model to describe at the molecular level the phase change of water as it turns from liquid water to steam at its boiling point.

5. When water boils, bubbles often appear and disappear upon reaching the surface of the boiling water. These bubbles are often made of:
    a. hydrogen molecules
    b. oxygen molecules
    c. hydrogen and oxygen molecules
    d. water molecules

6. Explain the reason behind your answer in question 5.

BLANK WORKSHEET

REVIEW QUESTIONS ON STRUCTURE AND PROPERTIES OF MATTER

1. In solids, the atoms or molecules are arranged such that they are close together. The reason the molecules stay close together is that the:

   a. attractive force between molecules is much greater than the kinetic energy of molecules

   b. kinetic energy of molecules is much greater than the attractive force between molecules

   c. attractive force between molecules is equal to the kinetic energy of molecules

   d. attractive force between molecules is much greater than the potential energy of molecules

2. In gases, the molecules move at random and collide with one another and walls of container. The reason gases behave this way is that the:

   a. attractive force between gas molecules is much greater than the kinetic energy of gas molecules

   b. kinetic energy of gas molecules is much greater than the attractive force between gas molecules

   c. attractive force between gas molecules is equal to the kinetic energy of gas molecules

   d. attractive force between gas molecules is much greater than the potential energy of gas molecules

3. If a single atom of reddish-brown copper metal is removed, will this copper atom also appear reddish-brown? Yes or No

   Explain

4. A syringe is filled half-way with air molecules and its tip capped to prevent air molecules from getting in or out of it. What will happen to the volume and mass of the air molecules when you apply force on the plunger?

   a. volume will decrease and mass of air will also decrease

   b. volume will decrease and mass of air will remain the same

   c. volume will decrease and mas of air will increase

   d. volume will increase and mass of air will also increase

LAWS OF NATURE AND THE ATOMIC THEORY

LEARNING OBJECTIVES

After this lesson, you should be able to:

- draw Thomson and Rutherford atomic models
- calculate the number of electrons or neutrons in an atom or ion
- state the properties of subatomic particles
- explain the term isotopes

WHAT NATURAL LAWS LED TO THE ATOMIC THEORY?

1.   Law of conservation of mass

2.   Law of constant composition

HOW DID DALTON'S ATOMIC THEORY EXPLAIN THESE LAWS

- All elements consist of small independent, indestructible and invisible particles called atoms.

- All atoms of the same element have equal mass. While atoms of different elements have different mass, and the mass of these atoms remain the same after physical or chemical change.

- Few atoms of different elements usually bond to form small particles. These small particles are identical in number and the types of atoms they contain, and a larger mass of a substance is simply an accumulation of these small particles.

WHAT IS DALTON'S ATOMIC MODEL?

DRAW THOMSON'S ATOMIC MODEL?

DRAW RUTHERFORD'S ATOMIC MODEL?

DRAW BOHR'S ATOMIC MODEL?

QUESTIONS ON ATOMIC THEORY

1. How did Thomson come to the conclusion that cathode-rays consist of particles that carry a negative charge?

2. How did Thomson come to the conclusion that the negatively charged particles were part of an atom?

3. How did Thomson come to the conclusion that all atoms consist of negatively charged particles.

4. How did Rutherford come to the conclusion that an atom consists of mostly empty space?

5. How did Rutherford come to the conclusion that an atom consists of a dense center, called the nucleus?

6. How did Rutherford come to the conclusion that the nucleus of an atom carry a positive charge?

WHAT ARE SUBATOMIC PARTICLES AND HOW ARE THESE PARTICLES RELATED?

Table 3: Properties of the proton, neutron and electron

Particle	Approximate mass	Relative charge	Location in atom	Symbol	Amount of space they occupy in an atom
Proton	1 (1.0072765 amu)	+1	Nucleus	p	A little
Neutron	1 (1.0086650 amu)	0	Nucleus	n	A little
Electron	0 (0.00054858 amu) So small its mass can be ignored	-1	Space around the nucleus	e⁻	A lot

Note: Exact mass values in parenthesis

What is **Atomic number (Z)**?

What is **Neutron number (N)**?

What is **Mass number (A)**?

Can an **atom** be **neutral**?

WHAT SYMBOLS ARE USED TO DESCRIBE AN ATOM?

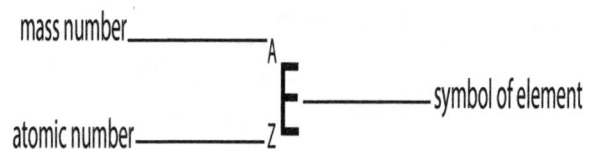

mass number _____ A

atomic number _____ Z E _____ symbol of element

FIND THE MISSING VALUES IN THE TABLE BELOW

| Chemical symbol | Number of | | | Mass number | Atomic number |
	Protons	Neutrons	Electrons		
$^{24}_{12}Mg$					
^{84}Kr					
	6	7			
		6	6		

Note. All the elements are neutral

SHOW WORK HERE

50

BLANK WORKSHEET

ATOMS

```
                          ┌──────────┐
                          │  Atoms   │
                          └──────────┘
                               │
                               │ can exist as
                               │
                ┌──────────────┴──────────────┐
          ┌──────────┐                   ┌──────────┐
          │          │                   │          │
          └──────────┘                   └──────────┘
               │                              │
         ┌─────┴─────┐                  ┌─────┴─────┐
   ┌────────┐   ┌────────┐        ┌────────┐   ┌────────┐
   │        │   │        │        │        │   │        │
   └────────┘   └────────┘        └────────┘   └────────┘
```

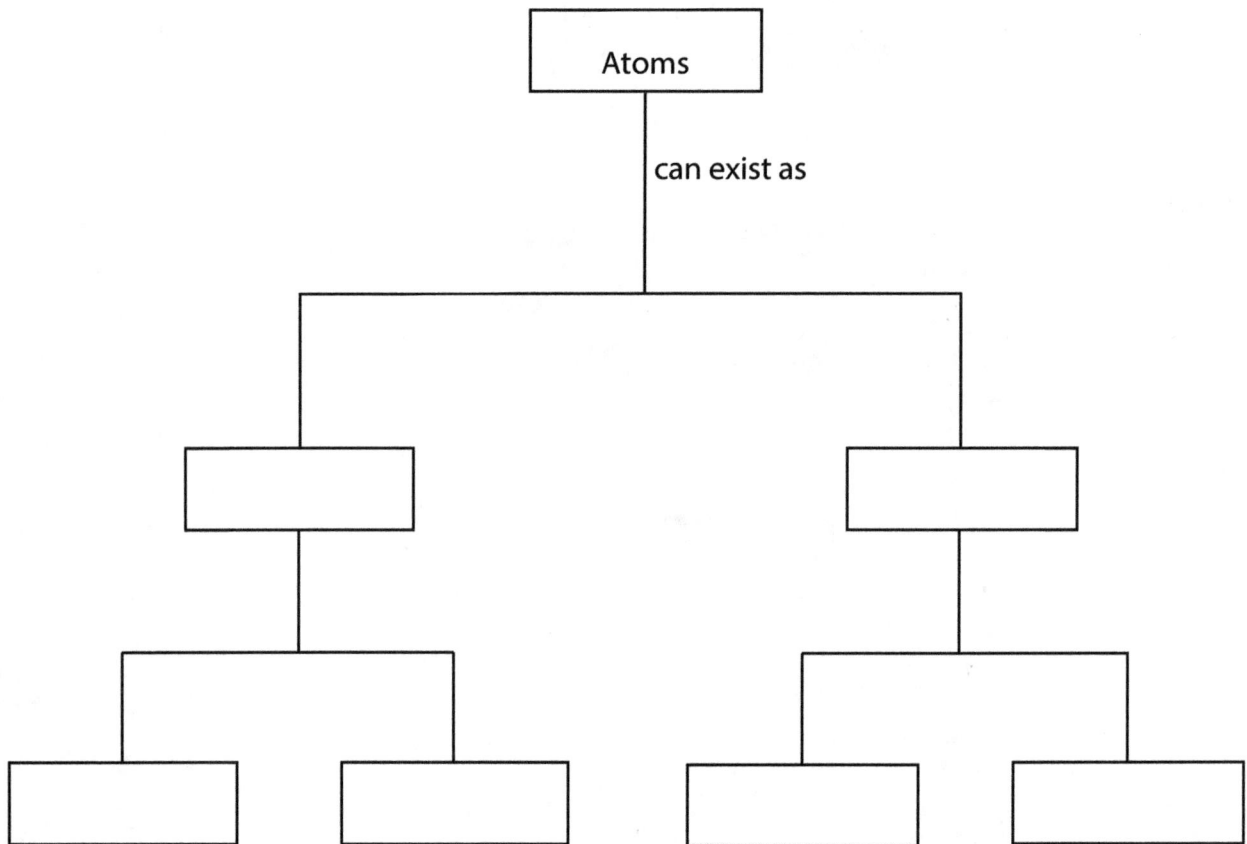

What is an **ISOTOPE?**

What is an **ION?**

What is a **CATION?**

What is an **ANION?**

WHAT SYMBOLS ARE USED TO DESCRIBE AN ION?

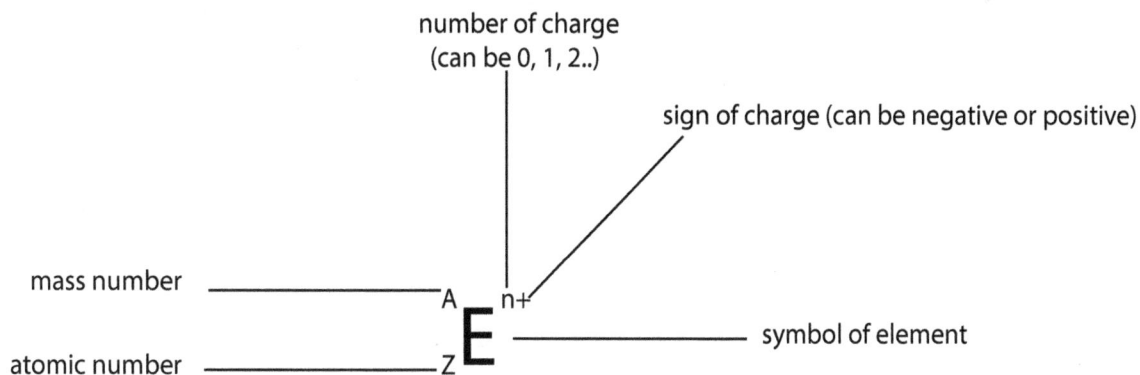

number of charge
(can be 0, 1, 2..)

sign of charge (can be negative or positive)

mass number $\longrightarrow$

atomic number $\longrightarrow$

$$_{Z}^{A}E^{n+}$$

symbol of element

Figure 1: Symbol of an element with charge, mass, and atomic number

$$_{12}^{24}Mg^{2+}$$ $\longrightarrow$ charge

Figure 2: Magnesium ion with mass and atomic number

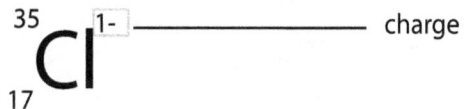

$$_{17}^{35}Cl^{1-}$$ $\longrightarrow$ charge

Figure 3: Chloride ion with mass and atomic number

For Mg^{2+} and Cl^- ions, calculate the number of

protons

electrons

neutrons

PERIODIC TABLE

1 — Group number																	18
1 H 1.01	2		Key:		Metal							13	14	15	16	17	2 He 4.00
3 Li 6.94	4 Be 9.01		atomic number Symbol atomic mass		Nonmetal							5 B 10.81	6 C 12.01	7 N 14.01	8 O 16.00	9 F 18.99	10 Ne 20.18
11 Na 22.99	12 Mg 24.31	3	4	5	6	7	8	9	10	11	12	13 Al 26.98	14 Si 28.09	15 P 30.97	16 S 32.06	17 Cl 35.45	18 Ar 39.95
19 K 39.10	20 Ca 40.08	21 Sc 44.96	22 Ti 47.88	23 V 50.94	24 Cr 52.00	25 Mn 54.94	26 Fe 55.85	27 Co 58.93	28 Ni 58.69	29 Cu 63.55	30 Zn 65.38	31 Ga 69.72	32 Ge 72.59	33 As 74.92	34 Se 78.96	35 Br 79.90	36 Kr 83.80
37 Rb 85.47	38 Sr 87.62	39 Y 88.91	40 Zr 91.22	41 Nb 92.91	42 Mo 95.94	43 Tc 98	44 Ru 101.07	45 Rh 102.91	46 Pd 106.42	47 Ag 107.88	48 Cd 112.41	49 In 114.82	50 Sn 118.69	51 Sb 121.75	52 Te 127.60	53 I 126.90	54 Xe 131.29
55 Cs 132.91	56 Ba 137.30	57-71 Lanthanoids	72 Hf 178.49	73 Ta 180.95	74 W 183.84	75 Re 186.21	76 Os 190.23	77 Ir 192.22	78 Pt 195.08	79 Au 196.97	80 Hg 200.59	81 Tl 204.38	82 Pb 207.2	83 Bi 208.98	84 Po	85 At	86 Rn
87 Fr	88 Ra	89-103 Actinoids	104 Rf	105 Db	106 Sg	107 Bh	108 Hs	109 Mt	110 Ds	111 Rg	112 Cn	113 Nh	114 Fl	115 Mc	116 Lv	117 Ts	118 Og

Metalloid

These rows go here

57 La 138.91	58 Ce 140.12	59 Pr 140.91	60 Nd 144.24	61 Pm 145	62 Sm 150.36	63 Eu 151.96	64 Gd 157.25	65 Tb 158.93	66 Dy 162.50	67 Ho 164.93	68 Er 167.26	69 Tm 168.93	70 Yb 173.04	71 Lu 174.97
89 Ac	90 Th 232.04	91 Pa 231.04	92 U 238.02	93 Np	94 Pu	95 Am	96 Cm	97 Bk	98 Cf	99 Es	100 Fm	101 Md	102 No	103 Lr

1. What is atomic mass (atomic weight) and is it the same as the mass number?

2. What is the periodic table?

3. How are the elements organized in the periodic table?

HOW ARE THE ELEMENTS ORGANIZED ON THE PERIODIC TABLE

Elements are arranged in vertical columns called **groups or families** and in horizontal rows called **periods**. The groups are numbered from **1** through **18**. And periods, from **1** through **7**. Elements in the same group share similar chemical properties and some physical properties.

- Elements in group **1** and **2** usually react with water to form **basic** solutions. As a result, **group 1** elements are called **alkali metals**, while group 2, **alkaline earth metals**. Group 1 elements all have **one electron** in their outermost shell **(highest occupied energy level)**. While **group 2** elements have **two electrons** in their highest occupied energy level.

- Elements in **Group 17** (group 7) are usually called the **halogens, which means salt forming in Greek**. They all have seven electrons in their highest occupied energy level.

- Elements in **group 18** (group 8) are usually called **noble** gases. Noble gases are highly **unreactive**. As a result, they are sometimes called the inert gases. Among them, only helium has two electrons in its highest occupied energy level, while the rest have eight electrons.

Elements can further be classified as **metals**, **nonmetals**, and **metalloids. Metalloids** or **semi-metals** have properties that are in-between those of metals and nonmetals.

Nonmetals can exist as gases, solids, or liquids. They tend to be brittle and do not conduct electricity and heat so well. Because of this behavior, they are usually called **insulators.** Nonmetals tend to form negative ions when they receive electrons from other atoms.

Except mercury, **metals** are usually solids. They tend to be shiny, ductile, malleable and can conduct electricity and heat so well. Because of this behavior, they are usually called **conductors.** Metals tend to form positive ions when they lose electrons to other atoms.

Depending on conditions, **metalloids** can behave as metals or nonmetals. They are neither good nor bad conductors of heat or electricity. Thus, they do not conduct heat or electricity as well as the metals, and they do not insulate as well as the nonmetals. Because of this middling behavior, they are usually called **semi-conductors**. Semi-conductors are used to make electronic chips in computers, hence the name **SILICON VALLEY** in California.

HOW TO DETERMINE ATOMIC MASS

LEARNING OBJECTIVES

After this lesson, you should be able to:

- apply weighted average to calculate atomic mass
- calculate molar mass
- calculate how many moles, atoms, or molecules are in a substance

Why are atomic masses unitless?

Atomic masses are unitless because they are **relative atomic masses. Relative** in this sense means one thing is compared to another. So, **relative atomic mass** means the mass of one atom is compared to the mass of another atom. **The atom to which other atoms are compared to is usually called the standard**. At present, an isotope of carbon called carbon-12 (C-12), which consists of 6 protons and 6 neutrons, is selected as the standard and assigned an atomic mass of exactly 12 amu, where amu stands for atomic mass units. Therefore, the mass of every other atom on the periodic table is determined by how light or heavy it is when compared to the mass of C-12.

What's percentage abundance? It is the proportion of atoms of an isotope in a sample of an element taken from the natural world. Percentage abundance is always reported as a percentage, which can be divided by 100 to get fractional abundance.

What instrument is used to measure relative masses? At present, **mass spectrometry** is the technique used to measure the relative masses of atoms. For example, a chemist measured a sample in a mass spectrometer and determined that the mass ratio of ^{28}Si (silicon-28) to ^{12}C (carbon-12) is: 2.33. Then, it follows that the mass of ^{28}Si = mass ratio of ($^{28}Si/^{12}C$) multiplied by mass of ^{12}C. Thus, mass of ^{28}Si = 2.33 x 12 amu = 27.98 amu.

HOW TO USE WEIGHTED AVERAGE TO CALCULATE ATOMIC MASS

If you examine the atomic masses on the periodic table, you will notice that the atomic masses are fractional. Why are they fractional? They are fractional because atoms exist as isotopes. And isotopes do not have the same mass and abundance in nature. As a result, to calculate the atomic mass of an element, we have to calculate how much each isotope contributes to the mass of the atom. To accomplish this, we usually use an approach called the weighted average. The weighted average takes into account the mass and natural abundance of each isotope. The weighted average also gives more **"weight"** to the values that appear more often.

Question

Use the table below to calculate the atomic mass of oxygen

Isotope	Mass, amu	Abundance, %
Oxygen-16	15.995	99.76
Oxygen-17	16.999	0.04
Oxygen-18	17.999	0.2

Solution

To calculate the atomic mass of oxygen, you must

- first multiply the mass of each isotope by its corresponding natural abundance. But, since the abundance is in %, you must divide each abundance value by 100.

- Sum the result to get the atomic mass of the element.

Thus, atomic mass of oxygen = 15.995 amu (99.76/100) + 16.999 amu (.04/100) + 17.999 amu (.2/100)

$$= 15.956612 \text{ amu} + 0.0067996 \text{ amu} + 0.035998 \text{ amu}$$

$$= 15.9994096 \text{ amu}$$

$$= 16.00 \text{ amu}.$$

Note that the abundance in percent always adds up to 100 %.

THE MOLE CONCEPT

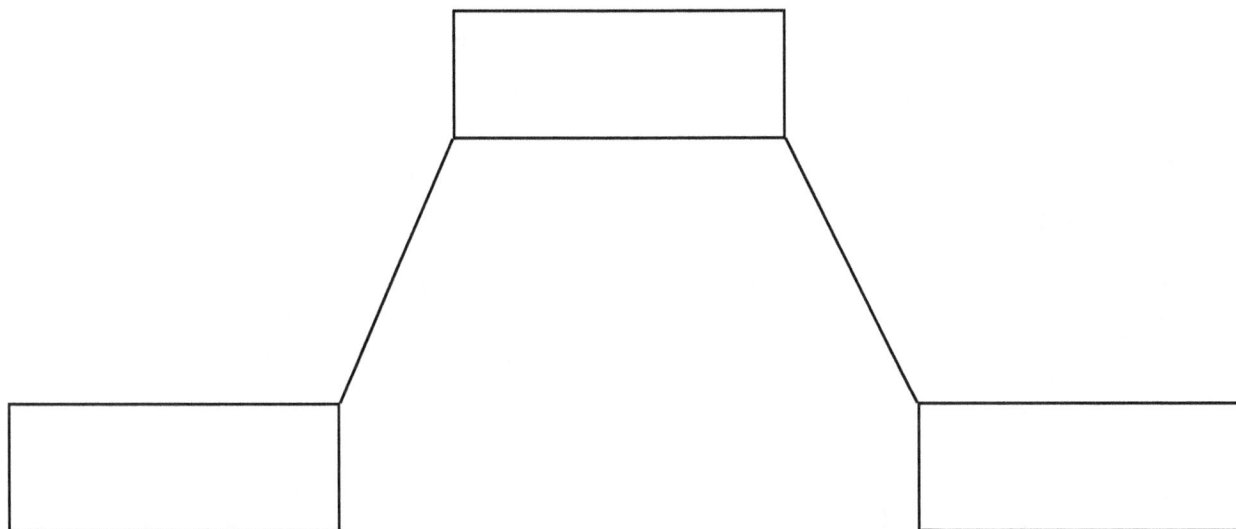

Why is the mole concept important in chemistry?

What's the Amount of Substance?

The **Amount of substance** is a quantity that measures the **size of a pile**(collection) of particles. These particles can be electrons, atoms, molecules, ions, or formula units. The amount of substance is an *SI* measured quantity that has a symbol *n*, and a base unit of *mole*, which is often written in a short way as *mol*.

What's the Mole?

The mole is the size of a pile of particles (**amount of substance)** that contains as many particles (electrons, atoms, molecules, ions, or formula units) as there are atoms in 12 grams of carbon-12 **(an isotope of carbon).**

What amount of substance can have the same number of particles as in 12g of carbon-12? The amount of substance that can have the same number of particles include:

1.   atomic mass of an element expressed in grams

2.   molecular mass of molecular compounds expressed in grams

3.   formula mass of ionic compounds expressed in grams

Why is it that 1 mole of any element has the same number of particles as in 1 mole of carbon-12?They do because of **Avogadro's hypothesis**. This hypothesis states that equal volumes of gases at the same temperature and pressure contain equal numbers of molecules. **How many particles (atoms, molecules, or ions) are there in 1 mole of substance?** There are 6.02×10^{22} particles in 1 mole of any substance. As you can see, this number is so huge that it has a special name called **Avogadro's number** in honor of Avogadro.

HOW TO MOVE FROM ATOMIC MASS (amu) TO GRAMS and TO GRAMS PER MOLE

Name of sub-stance	Atomic mass of substance (amu)	1 mole of substance	Mass of 1 mole of substance in grams, g	Molar mass of substance in grams per mole (g/mol)
Hydrogen (H)				
Oxygen (O)				

How to calculate the molar mass of a compound

Calculate the molar mass of

1. CO_2

2. $Al(OH)_3$

HOW TO RELATE MASS, MOLE, AND AVOGADRO'S NUMBER

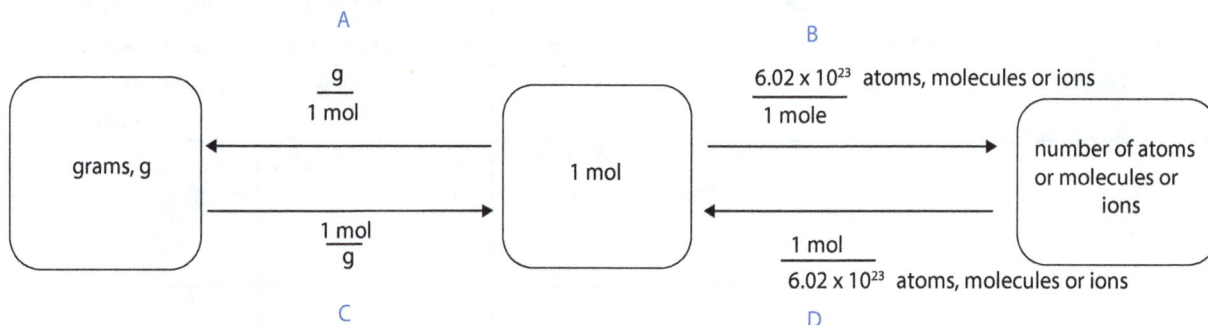

A

$$\frac{g}{1\ mol}$$

B

$$\frac{6.02 \times 10^{23}\ \text{atoms, molecules or ions}}{1\ mole}$$

| grams, g | | 1 mol | | number of atoms or molecules or ions |

C

$$\frac{1\ mol}{g}$$

D

$$\frac{1\ mol}{6.02 \times 10^{23}\ \text{atoms, molecules or ions}}$$

1.   To go from mol to grams, g, use the conversion factor A

2.   To go from grams, g to mol, use the conversion factor C

3.   To go from mol to number of atoms, molecules or ions, use conversion factor B

4.   To go from number of atoms, molecules, or ions to mol, use the conversion factor D

5.   To go from grams to number of atoms, you have to first use conversion factor A and then B

6.   To go from number of atoms to grams, you have to first use conversion factor D and then C

Pay attention to direction of arrows

BLANK WORKSHEET

MOLE CONCEPT PRACTICE PROBLEMS

1.

    a. Calculate the molar mass of sulfuric acid (H_2SO_4)

    b. If you have 5.0 g of sulfuric acid, how many moles of sulfuric acid are present in the sample?

    c. If you have 5.0 g of sulfuric acid, how many molecules of sulfuric acid are present in the sample?

    d. Use your answer in **question c** to convert the molecules of sulfuric acid back to grams of sulfuric acid

    e. Why is the mole concept useful to chemists?

REVIEW ON ATOMIC NUMBER, MASS NUMBER, AND NUMBER OF NEUTRONS

Calculate the number of protons, electrons, and neutrons for the following elements

[1]Hydrogen							[4]Helium
# of p:							# of p:
# of e:							# of e:
# of n:							# of n:
[7]Lithium	[9]Beryllium	[11]Boron	[12]Carbon	[14]Nitrogen	[16]Oxygen	[19]Fluorine	[20]Neon
# of p:	# of p:	# of p:	# of p:	# of p:	# of p:	# of p:	# of p:
# of e:	# of e:	# of e:	# of e:	# of e:	# of e:	# of e:	# of e:
# of n:	# of n:	# of n:	# of n:	# of n:	# of n:	# of n:	# of n:
[23]Sodium	[24]Magne-sium	[27]Alumi-num	[28]Silicon	[31]Phos-phorus	[32]Sulfur	[35]Chlorine	[40]Argon
# of p:	# of p:	# of p:	# of p:	# of p:	# of p:	# of p:	# of p:
# of e:	# of e:	# of e:	# of e:	# of e:	# of e:	# of e:	# of e:
# of n:	# of n:	# of n:	# of n:	# of n:	# of n:	# of n:	# of n:
[39]Potassi-um	[40]Calcium						
# of p:	# of p:						
# of e:	# of e:						
# of n:	# of n:						

HOW TO CREATE AN ELECTRON MAP FOR AN ATOM

LEARNING OBJECTIVES

After this lesson, you should be able to:

- write electron configurations for simple elments and their ions
- use the electron configurations to determine the number of valence electrons
- explain why atomic size increase down a group but decrease across a period
- explain why an ion is smaller or larger than its neutral atom

Bohr's Atomic model

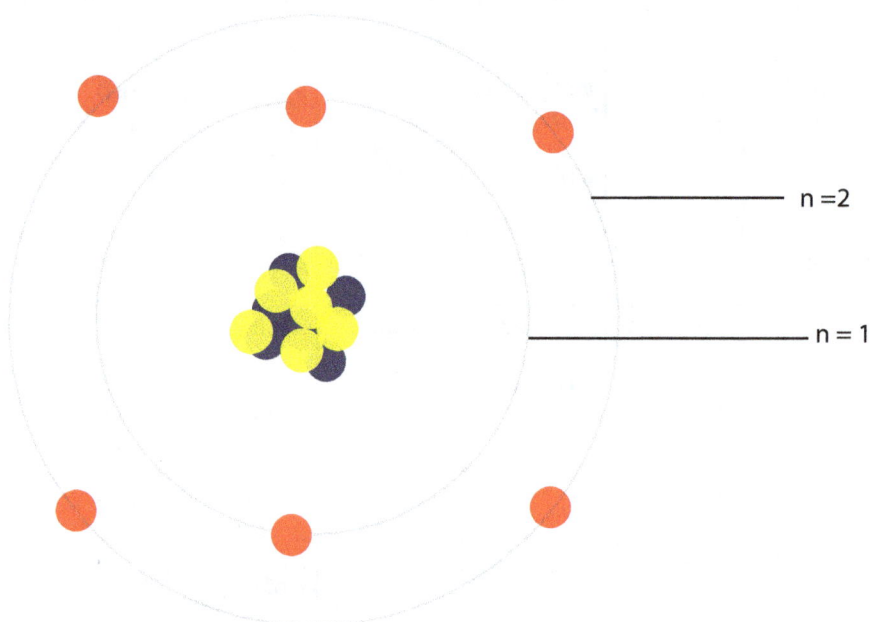

n = 1, 2, 3, ... where n = energy level

Each shell or energy level can hold a maximum of **$2n^2$ electrons**, where n is the energy level (principal quantum number of the shell). When n = 1, the maximum number of electrons the shell can hold is: $2(1)^2 = 2$ electrons. Electrons are filled starting from the lowest energy level and on. That is from n =1, 2, 3, and on.

HOW TO USE BOHR'S ATOMIC MODEL TO WRITE ELECTRON CONFIGURATIONS

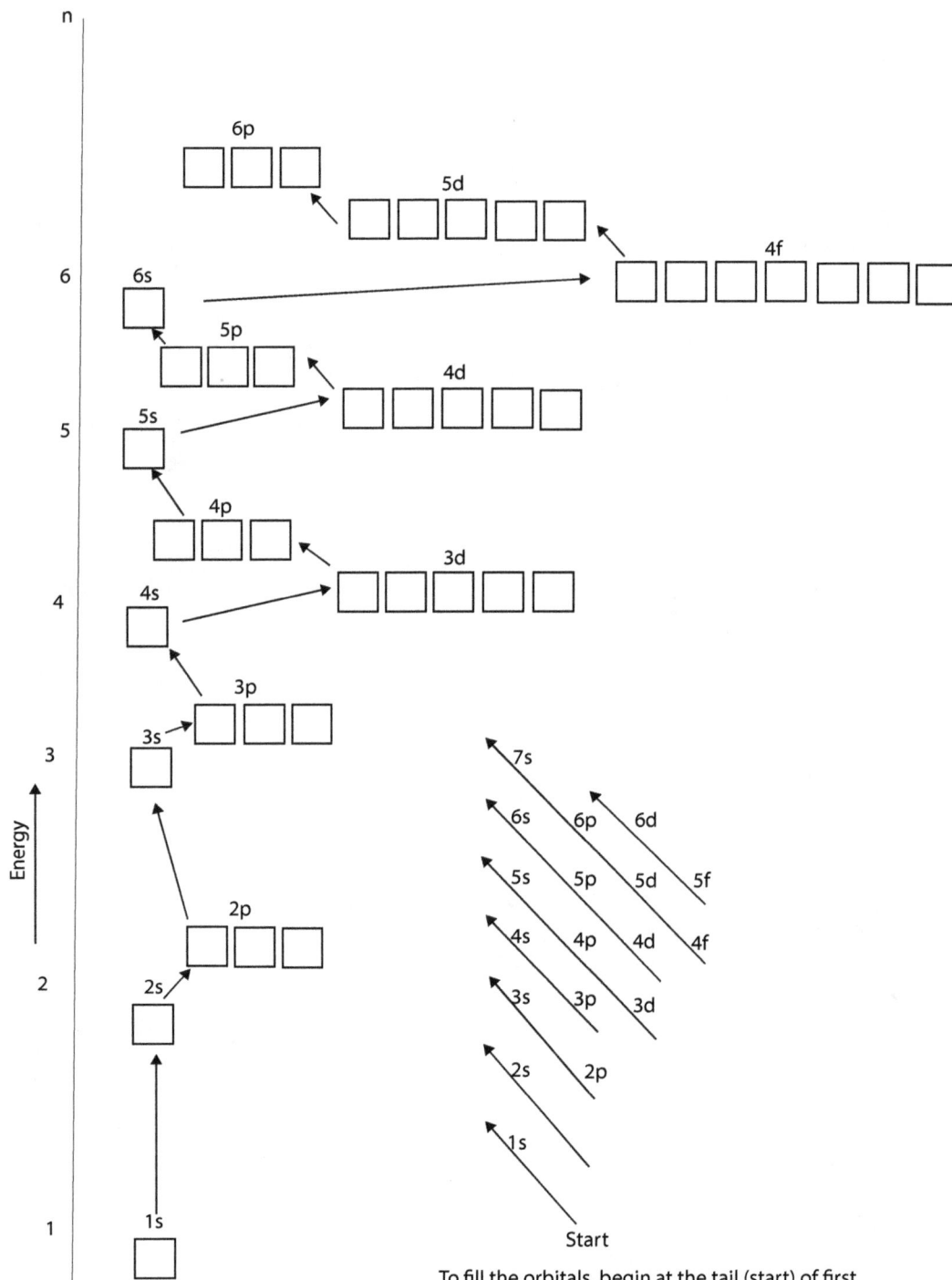

n

Energy

6p

5d

4f

6s

5p

4d

5s

4p

3d

4s

3p

3s

2p

2s

1s

7s

6s   6p   6d

5s   5p   5d   5f

4s   4p   4d   4f

3s   3p   3d

2s   2p

1s

Start

To fill the orbitals, begin at the tail (start) of first
arrow, and then go to the head of same arrow.
Repeat the same moves for the next arrow.

TYPES OF ORBITALS

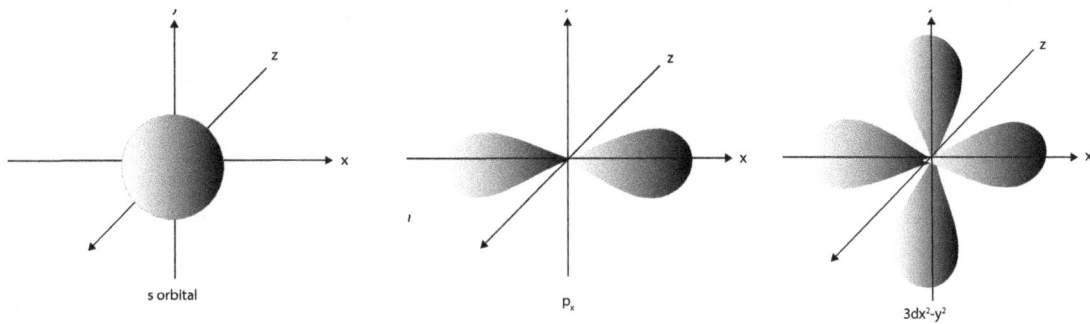

s orbital

p_x

$3dx^2-y^2$

THREE ORBITALS PUT TOGETHER

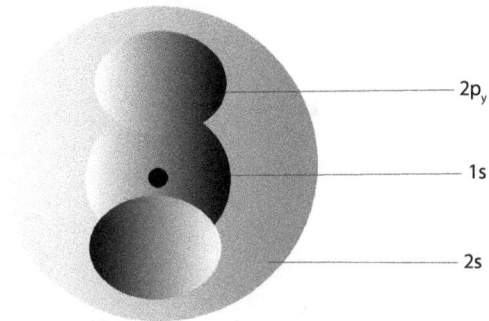

$2p_y$

1s

2s

ELECTRON SPINS

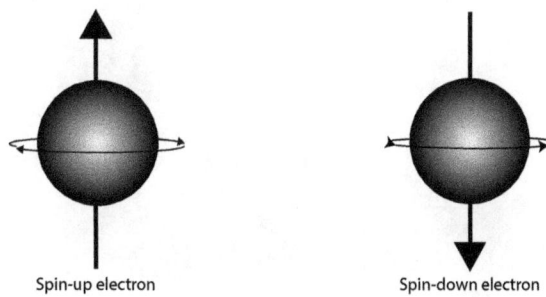

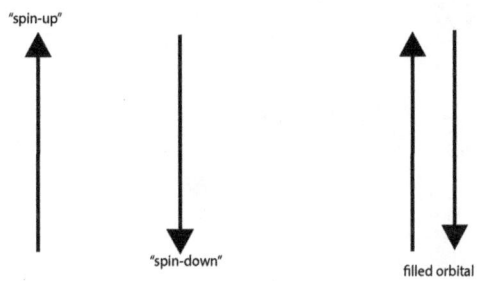

Spin-up electron

Spin-down electron

"spin-up"

"spin-down"

filled orbital

RULES FOR FILLING ATOMIC ORBITALS WITH ELECTRONS

RULE 1

RULE 2

RULE 3

RULE 4

Orbital diagram

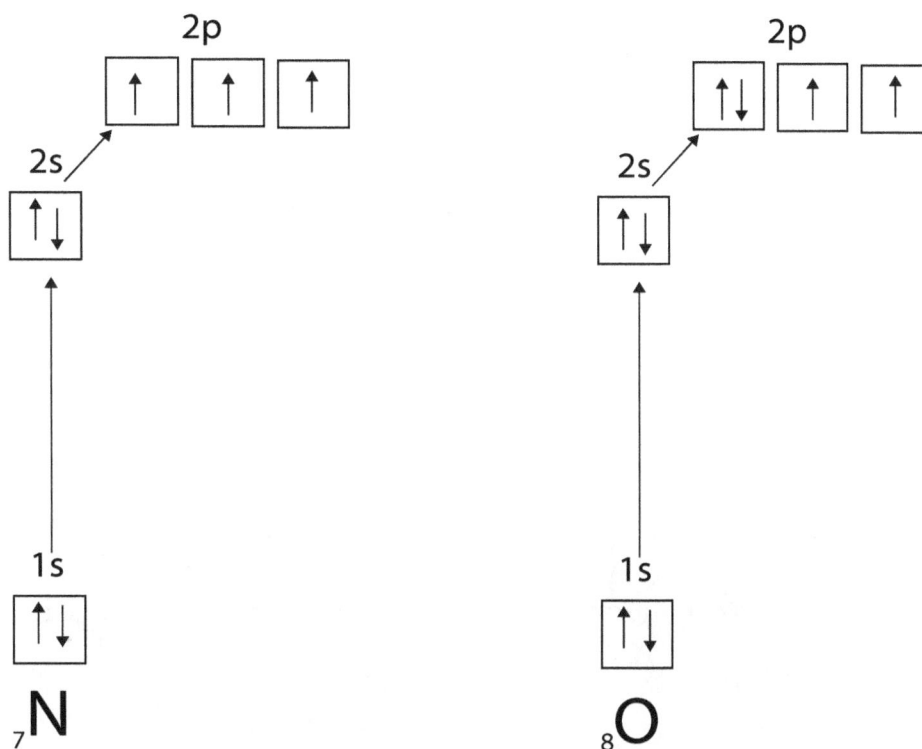

2p

2s

1s

$_7$N

2p

2s

1s

$_8$O

Electron configuration of Nitrogen (N)

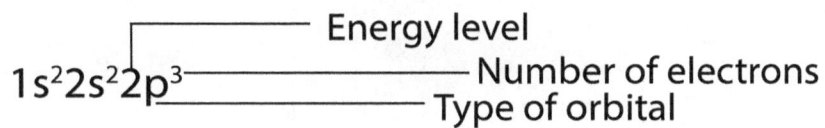

Energy level

Number of electrons

Type of orbital

$1s^2 2s^2 2p^3$

Electron configuration of Oxygen (O)

$1s^2 2s^2 2p^4$

QUESTIONS ON ELECTRON CONFIGURATIONS

1. Use the orbital filling diagram to write the electron configuration for:

   • Cl

   • Cl⁻

   • Na

   • Na⁺

2. From the electron configurations, determine the valenece electrons for

   • Cl

   • Na

3. From the electron configurations, determine the number of unpaired electrons for

   • Cl

   • Cl⁻

   • Na

   • Na⁺

Write and use the electron configuration to identify the number of valence electrons

Element	Electron configuration	Valence eletrons	Group number
Li			
Na			
Be			
Mg			
B			
Al			
C			
Si			
N			
P			
O			
S			
F			
Cl			
Ne			
Ar			

Question

For main group elements, what's the relationship between their valence electrons and their group number?

HOW ATOMIC SIZE VARY ALONG A GROUP OR PERIOD FOR MAIN GROUP ELEMENTS

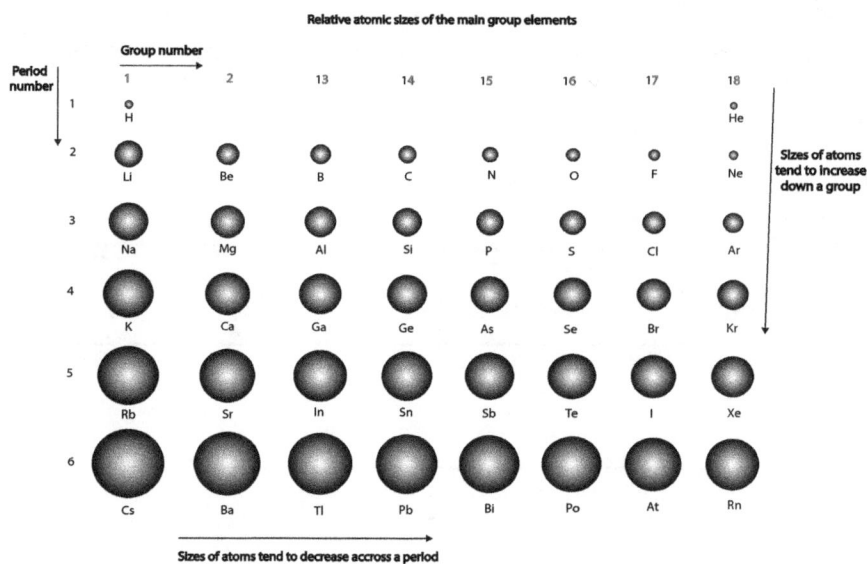

Figure 2

Question

1. From figure 2 above, why does **atomic size** increase **down** a **group,** but **decrease** across a **period**.

Neutral atom

Ion

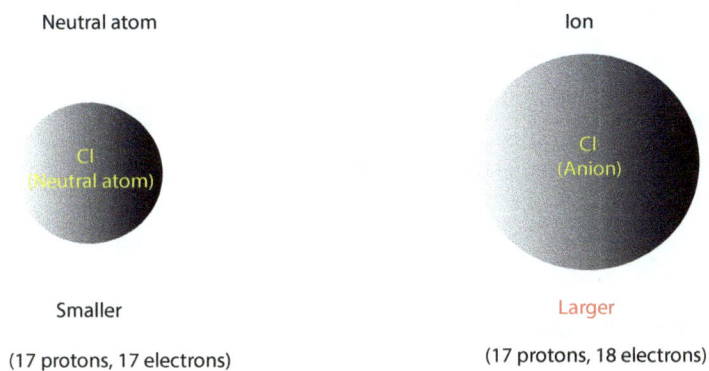

Cl
(Neutral atom)

Cl
(Anion)

Figure 3

Smaller

Larger

(17 protons, 17 electrons)

(17 protons, 18 electrons)

Question

3.   From figure 3 above, why is it that the negative ion(anion) is larger than its neutral atom?

Neutral atom

Ion

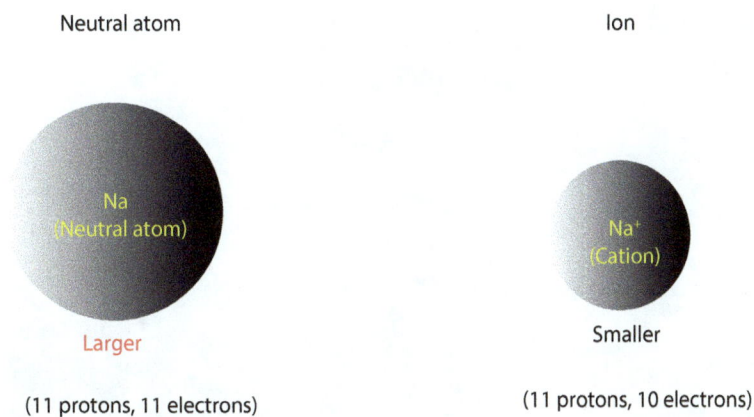

Na
(Neutral atom)

Na⁺
(Cation)

Figure 4

Larger

Smaller

(11 protons, 11 electrons)

(11 protons, 10 electrons)

Question

2.   From figure 4 above, why is it that the positive ion(cation) is smaller than its neutral atom?

LEWIS DOT STRUCTURES

LEARNING OBJECTIVES

After this lesson, you should be able to:

- draw Lewis dot structures for simple atoms and molecules
- define the octet rule

Draw Lewis dot structures for the following atoms

H

LI

Be

B

C

N

O

F

Ne

What is the octet rule? It states that atoms need 8 valence electrons in order to become stable.

What is one exception to the octet rule? Hydrogen only need two (duet) valence electrons to become stabe.

CHEMICAL BONDING

LEARNING OBJECTIVES

After this lesson, you should be able to:

- describe covalent and ionic bond
- describe the difference between intermolecular and intramolecular forces
- defined chemical bonding
- describe the physical properties of ionic and covalent compounds

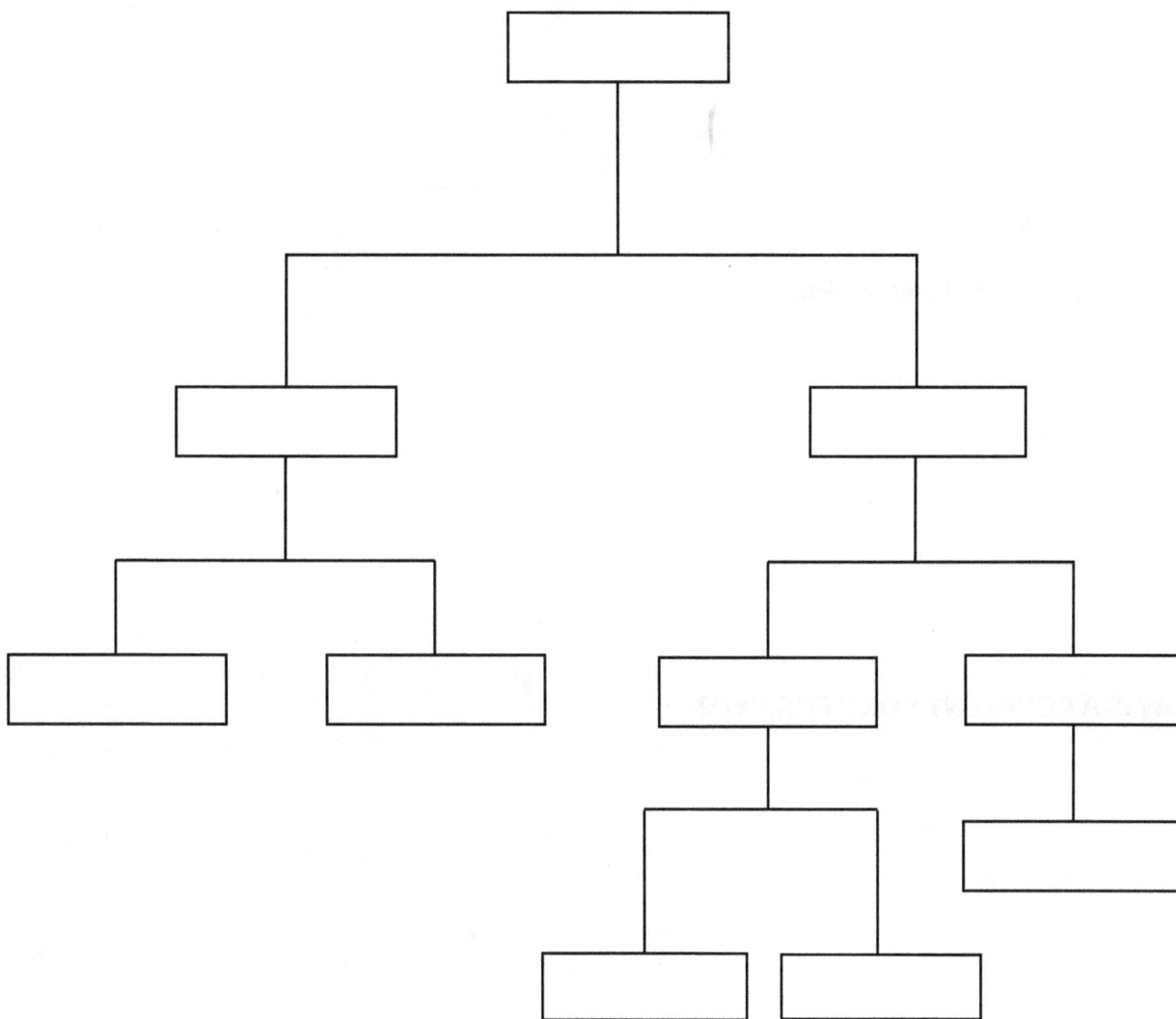

WHY DO ATOMS BOND?

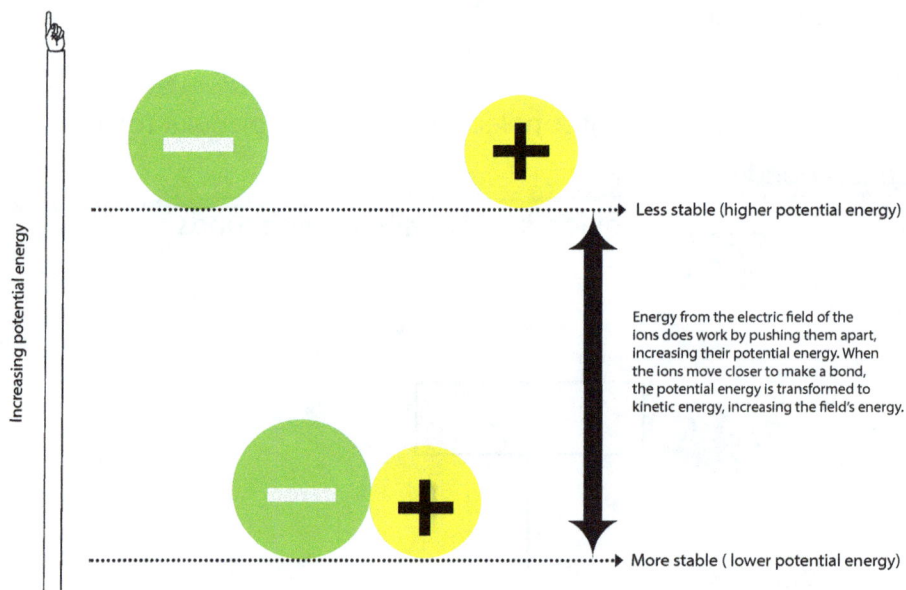

Increasing potential energy

Less stable (higher potential energy)

Energy from the electric field of the ions does work by pushing them apart, increasing their potential energy. When the ions move closer to make a bond, the potential energy is transformed to kinetic energy, increasing the field's energy.

More stable (lower potential energy)

HOW IS AN IONIC BOND FORMED?

HOW IS A COVALENT BOND FORMED?

How do hydrogen atoms form covalent bond in hydrogen molecule (H_2)?

DRAW LEWIS DOT STRUCTURES TO SHOW HOW THESE MOLECULES FORM COVALENT BONDS

1. N_2

2. NH_3

3. H_2O

4. CH_4

5. CH_2CH_2

6. PCl_3

How do sodium (Na) and chlorine (Cl) atoms form the ionic bond in sodium chloride?

HOW TO NAME COVALENT COMPOUNDS

HOW TO NAME IONIC COMPOUNDS

IDENTIFY THE TYPE OF COMPOUND AND NAME IT

Compound	Type of compound	Name

1. $AlCl_3$

2. PBr_3

3. Na_3PO_4

4. Mg_3N_2

5. N_2O

6. CBr_4

7. SCl_2

8. P_2O_5

9. KI

10. CaF_2

___

11. In a chemical compound, where is chemical energy stored? Explain

12. Is energy always released when atoms or molecules interact to form bonds? Explain

13. Is energy always required (put in) in order to break bonds? Explain

EXPERIMENT 4: PHYSICAL PROPERTIES OF IOIC AND COVALENT COMPOUNDS

Table 1: Physical properties of ionic and covalent compounds

Compound	Texture & color of compound	Melting point	Dissolves in DI water	Conductivity when solid	dissolved in DI H2O
Calcium Chloride (CaCl$_2$)		772°C			
Sodium Chloride (NaCl)		801°C			
Sucrose (C$_{12}$H$_{22}$O$_{11}$)		186 °C			
Stearic Acid (C$_{18}$H$_{36}$O$_2$)		69 °C			

Note: *DI* means Deionized water

POST-LAB QUESTIONS

1. Use ability of compound to dissolve in water to classify the compounds into 2 groups

   Dissolves in water   **Does not dissolve in water**

2. Use conductivity to classify the compounds into 2 groups.

   Conduct when dissolved   Does not conduct when dissolved

3. Use the physical property of melting point to classify ionic and covalent compounds into two groups.

4. From your data, describe the general physical properties of ionic compounds.

5. From your data, describe the general physical properties of covalent compounds.

6. From your data, which physical property is least helpful in separating the compounds into ionic or covalent? Why?

7. List the physical properties you would predict for the following compounds:
   glucose (C$_6$H$_{12}$O$_6$), Sodium carbonate (Na$_2$CO$_3$) and Potassium sulfate (K$_2$SO$_4$).

BIANK WORKSHEET

CHEMICAL REACTIONS

LEARNING OBJECTIVES

After this lesson, you should be able to:

- describe signs and properties of chemical reactions
- balance chemical equations and verify that mass is conserved
- describe types of chemical reactions
- write basic chemical equations

What's a chemical reaction?

A chemical reaction is the same as a chemical change. And it occurs when chemicals, called reactants react to form new chemicals, called products. The new chemicals usually have their own properties. And these properties differ from the properties of the reactants.

Characteristics of chemical reactions

- mass is always conserved
- new chemicals with unique properties are produced (composition changes)

What're the signs of chemical reactions?

Types of reactions based on how atoms behave when they react with each other

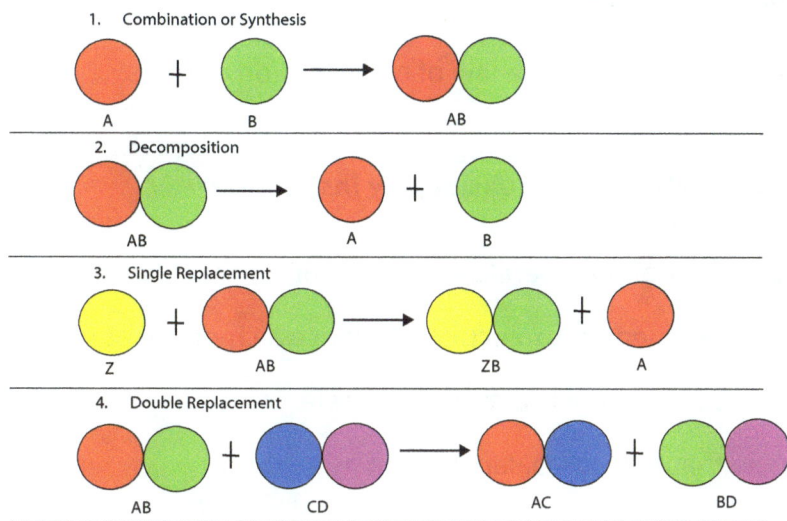

1. **Combination or Synthesis**

A + B → AB

2. **Decomposition**

AB → A + B

3. **Single Replacement**

Z + AB → ZB + A

4. **Double Replacement**

AB + CD → AC + BD

What's a chemical equation?

A chemical equation is a short way of describing a chemical reaction. Here is a chemical equation that describes at the symbolic and particulate level the reaction between sodium and chlorine.

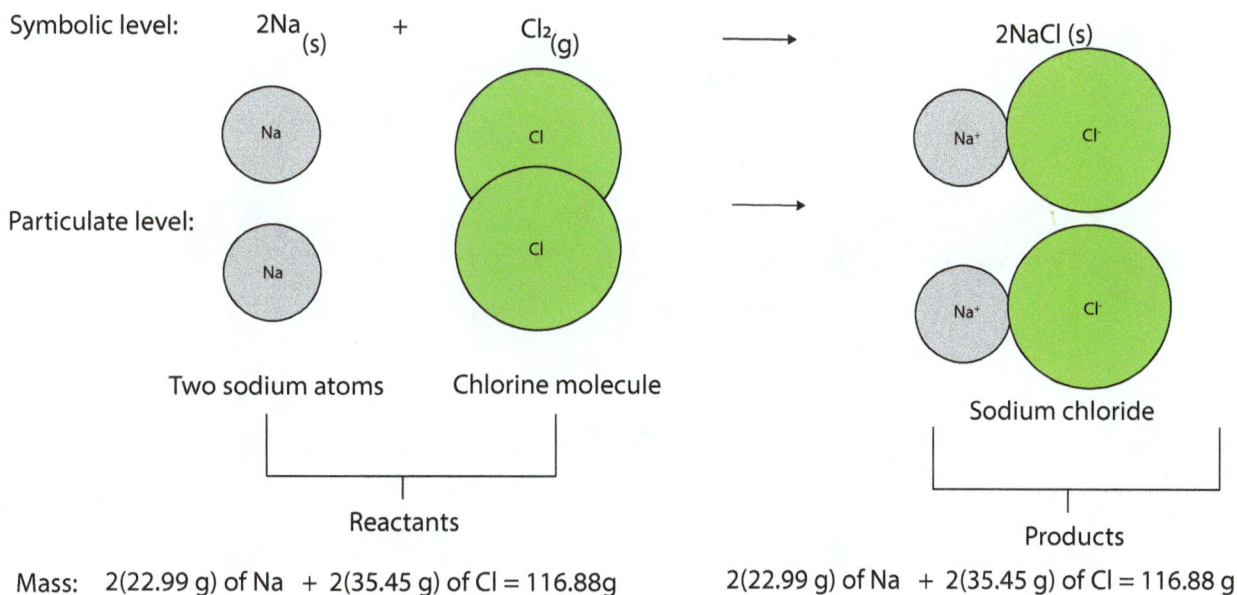

Symbolic level: $2Na_{(s)}$ + $Cl_{2(g)}$ → $2NaCl_{(s)}$

Particulate level:

| Na | Cl | Na⁺ Cl⁻ |

Two sodium atoms   Chlorine molecule   Sodium chloride

Reactants   Products

Mass: 2(22.99 g) of Na + 2(35.45 g) of Cl = 116.88g   2(22.99 g) of Na + 2(35.45 g) of Cl = 116.88 g

Law of conservation of mass: Mass of reactants is equal to mass of products

HOW TO BALANCE CHEMICAL EQUATIONS

Why should chemical equations be balanced?

We must balance chemical equations because of the law of conservation of mass.

Basic rules you can apply to balance chemical equations by inspection

- Place numbers as coefficients in front of the chemical formulas or symbols of elements.

- Never change the subscript of an element or molecule to balance a chemical equation. When you do, you will change the identity of the molecule or element.

- Keep track of polyatomic ions so that you can balance them as a group.

- Balance elements that stand alone last.

Example

BALANCE THE FOLLOWING CHEMICAL EQUATIONS BY INSPECTION

1. $SO_2 + O_2 \ ---> \ SO_3$

2. $KClO_3 \ ---> \ KCl + O_2$

3. $CaH_2 + H_2O \ ---> \ Ca(OH)_2 + H_2$

4. $Na + O_2 \ ---> \ Na_2O$

5. $AgNO_3 + MgCl_2 \ ---> \ AgCl + Mg(NO_3)_2$

6. $AlBr_3 + K_2SO_4 \ ---> \ KBr + Al_2(SO_4)_3$

7. $CH_4 + O_2 \ ---> \ CO_2 + H_2O$

8. $FeCl_3 + NaOH \ ---> \ Fe(OH)_3 + NaCl$

9. $P + O_2 \ ---> \ P_2O_5$

10. $Al + H_3PO_4 \ ---> \ H_2 + AlPO_4$

IDENTIFY THE TYPE OF REACTION

1. $H_2 + O_2 \ ---> \ H_2O$

2. $Zn + H_2SO_4 \ ---> \ ZnSO_4 + H_2$

3. $AgNO_3 + NaCl \ ---> \ AgCl + NaNO_3$

DO THESE EQUATIONS DESCRIBE A CHEMICAL REACTION?

1. $H_2O(l) \ ---> \ H_2O(g)$

2. $O_2(g) + O(g) \ ---> \ O_3(g)$

3. $H_2(g) + HCl(aq) \ ---> \ H_2(g) + HI(g)$

EXPERIMENT 5: EXPLORING THE SIGNS AND TYPES OF CHEMICAL REACTIONS

1. Put about 1.00 g of $CuSO_4.5H_2O(s)$ into your test tube. Then use a test-tube holder to gently hold your test tube while you heat substance in flame.

    a. Why did substance in test tube change color?

    b. Write a balanced chemical equation for the reaction.

    c. What type of reaction is this?

2. After your test tube cools, add 3 drops of water to compound.

    a. Why did substance in test tube change color?

    b. Write a balanced chemical equation for the reaction.

    c. What type of reaction is this?

3. Put about 5.0 mL of HCl (aq) solution into your test tube. Then add a strip of Al(s) foil into the tube.

    a. Describe your observations, and what in your observations shows that it is a a chemical reaction.

    b. Write a balanced chemical equation for the reaction.

    c. Draw a particle model for the chemical equation.

    d. What type of reaction is this?

4. Put 4 drops of Na_2SO_4 (aq) into your test tube, then add another 4 drops of $BaCl_2$ (aq).

    a. Describe your observations.

    b. Write a balanced chemical equation for the reaction.

    c. What type of reaction is this?

BIANK WORKSHEET

Some Ions of Main Group Elements

1 ——————— Group number   18

H^+	2		13	14	15	16	17	
Li^+					N^{3-}	O^{2-}	F^-	
Na^+	Mg^{2+}		Al^{3+}		P^{3-}	S^{2-}	Cl^-	
K^+	Ca^{2+}						Br^-	
Rb^+	Sr^{2+}						I^-	
Cs^+	Ba^{2+}							

Note that a salt of a +1 cation or -1 anion is likely to dissolve in water

HOW TO USE BALANCED CHEMICAL EQUATION TO CALCULATE THE AMOUNT OF REACTANTS YOU WILL NEED OR AMOUNT OF PRODUCTS YOU CAN MAKE (STOICHIOMETRY)

LEARNING OBJECTIVES

After this lesson, you should be able to:

- write mole-mole ratio
- calculate moles of reactant or product
- verify law of conservation of mass
- define limiting reactant

How to make cheesecake

Here is a recipe to make cheesecake:

1 cheesecake
_____

8 blocks cream cheese

4 cups sugar

12 eggs

1 cup of flour

Use the recipe to answer the following questions

1.   You have only 6 cups of sugar (assume you have the other ingredients in excess). How many cakes can you make?

2.   If you use 6 cups of sugar (assume you have the other ingredients in excess), how many eggs will you need?

3.   If you have 50 cups of sugar, 50 blocks of cream cheese, 5 eggs, and 4 cups of flour how many cheesecakes can you make?

HOW TO USE A BALANCED CHEMICAL EQUATION TO WRITE MOLE-MOLE RATIO

Use the following balanced chemical equation to write mole-mole ratio

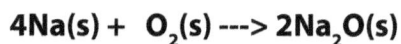

$$4Na(s) + O_2(s) ---> 2Na_2O(s)$$

Write the mole-mole ratio for:
- a.   Na and O_2
- b.   Na and Na_2O
- c. O_2 and Na_2O

Use the equation below to calculate how many moles of sulfur are needed to react with 1.42 moles of Fe.

$$2Fe(s) + 3S(s) ---> Fe_2S_3(s)$$

HOW TO VERIFY LAW OF CONSERVATION OF MASS

Balance the following chemical equation and use it to verify the law of conservation of mass

$$CH_4(g) + O_2(g) ---> CO_2(g) + H_2O(g)$$

1.   Calculate the total mass of reactants.

2.   Calculate the total mass of products.

3.   Is the mass of reactants equal to the mass of products? Explain.

SOLUTIONS

LEARNING OBJECTIVES

After this lesson, you should be able to:

- define a solution
- explain why a solute dissolves in a solvent
- prepare a solution and calculate its concentration in mass percent
- explain why some solutions conduct electricity, while others don't

What's a solution?

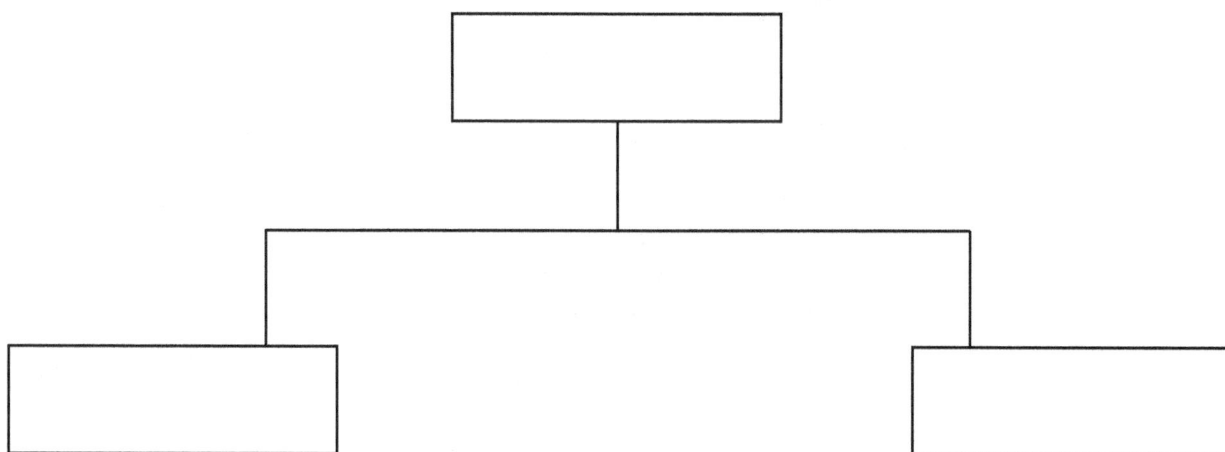

```
                    ┌─────────────────────┐
                    │                     │
                    │                     │
                    └──────────┬──────────┘
                               │
           ┌───────────────────┴───────────────────┐
           │                                        │
  ┌────────────────┐                      ┌────────────────┐
  │                │                      │                │
  │                │                      │                │
  └────────────────┘                      └────────────────┘
```

What's a solute?

What's a solvent?

Types of solutions

Solution type	Solute	Solvent	Example
Gas solution	Gas	Gas	Air (mixture of gases)
Liquid solution	Solid	Liquid	Sugar and water
	Liquid	Liquid	beer (water and ethanol)
	Gas	Liquid	Soda (carbon dioxide and water)
Solid solution	Solid	Solid	Brass (copper and zinc)

95

WHAT FACTORS DETERMINE WHETHER SOLUTE WILL DISSOLVE IN A SOLVENT?

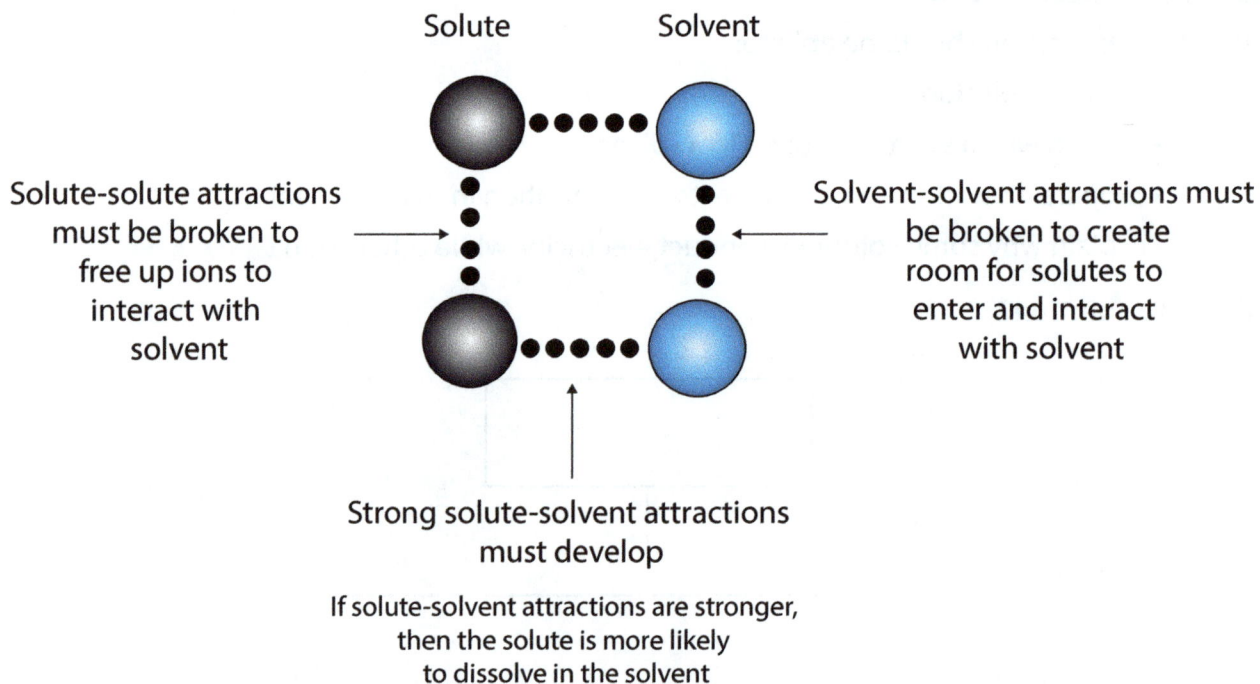

Solute   Solvent

Solute-solute attractions must be broken to free up ions to interact with solvent

Solvent-solvent attractions must be broken to create room for solutes to enter and interact with solvent

Strong solute-solvent attractions must develop

If solute-solvent attractions are stronger, then the solute is more likely to dissolve in the solvent

These factors can be summarized simply as "**LIKE DISSOLVES LIKE**."

LIKE DISSOLVES LIKE means:

1.

2.

HOW DOES SODIUM CHLORIDE (NaCl) DISSOLVE IN WATER (H_2O)?

Here is a molecular model that shows sodium chloride dissolved in water

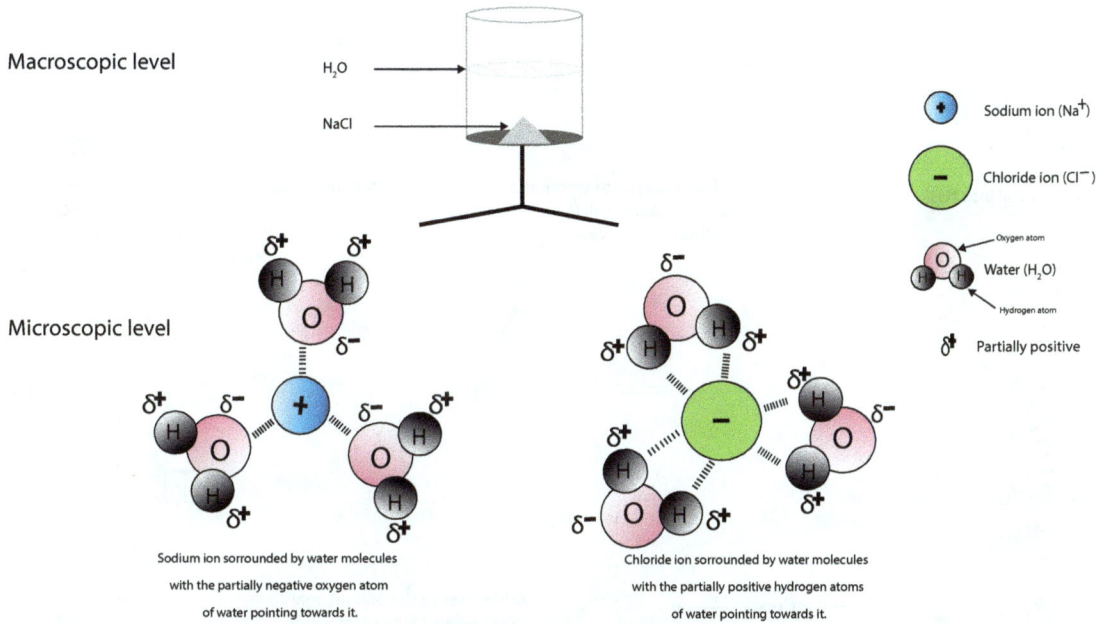

Macroscopic level

H_2O

NaCl

Microscopic level

Sodium ion (Na^+)

Chloride ion (Cl^-)

Oxygen atom

Water (H_2O)

Hydrogen atom

δ^+ Partially positive

Sodium ion sorrounded by water molecules
with the partially negative oxygen atom
of water pointing towards it.

Chloride ion sorrounded by water molecules
with the partially positive hydrogen atoms
of water pointing towards it.

Recall that:

    solute is salt (NaCl)

    Solvent is water (H_2O)

Explain how sodium chloride dissolves in water.

ACTIVITY 4.0: HOW TO PREPARE SOLUTION AND CALCULATE ITS CONCENTRATION

The picture below shows how to prepare solution. Use it as a guide to learn how to prepare and calculate the concentration of sugar solution

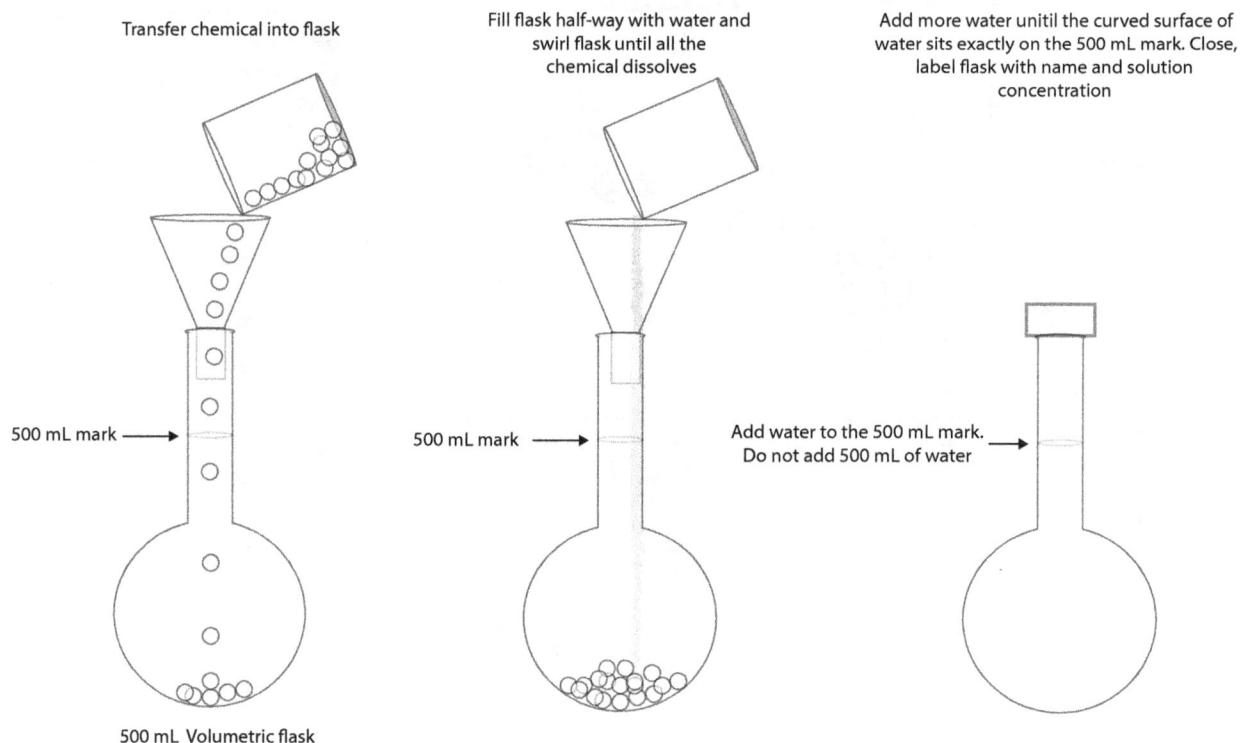

Transfer chemical into flask

Fill flask half-way with water and swirl flask until all the chemical dissolves

Add more water unitil the curved surface of water sits exactly on the 500 mL mark. Close, label flask with name and solution concentration

500 mL mark →

500 mL mark →

Add water to the 500 mL mark. Do not add 500 mL of water →

500 mL Volumetric flask

Steps

1. Weigh your empty flask, and record its mass in grams

2. Weigh between 2 to 6 grams of sugar. Record mass in grams

3. Transfer sugar into flask

4. Add enough water to dissolve sugar, then top it up to the mark

5. Weigh your flask with solution, and record its mass in grams

HOW TO CALCULATE SOLUTION CONCENTRATION

What is solution concentration?

What is solution concentration in mass percent?

What is solution concentration in volume percent?

What is solution concentration in moles per liter?

BLANK WORKSHEET

ACIDS AND BASES

LEARNING OBJECTIVES

After this lesson, you should be able to:

- define an acid and a base according to Arrhenius
- use particle models to describe strong acid or base and weak acid or base
- describe the pH scale and use it to identify an acid, base or neutral substance

General properties of acids and bases

At the macroscopic level, acids

- taste sour
- turns blue litmus paper red
- reacts with metals to produce hydrogen gas
- reacts with carbonates to produce carbon dioxide
- neutralize bases in an acid-base reaction to produce salt and water

At the macroscopic level, bases

- taste bitter
- produce aqueous solutions that feels slippery
- turns red litmus paper blue
- neutralize acids in an acid-base reaction to produce salt and water

At the molecular level, how do acids and bases behave?

To understand how acids and bases behave at the molecular level, we must explore why some aqueous solutions conduct electricity, while others do not. From our observations, we will logically explain how an acid and base behave at the molecular level.

HOW TO USE CONDUCTIVITY TO CLASSIFY AQUEOUS SOLUTIONS

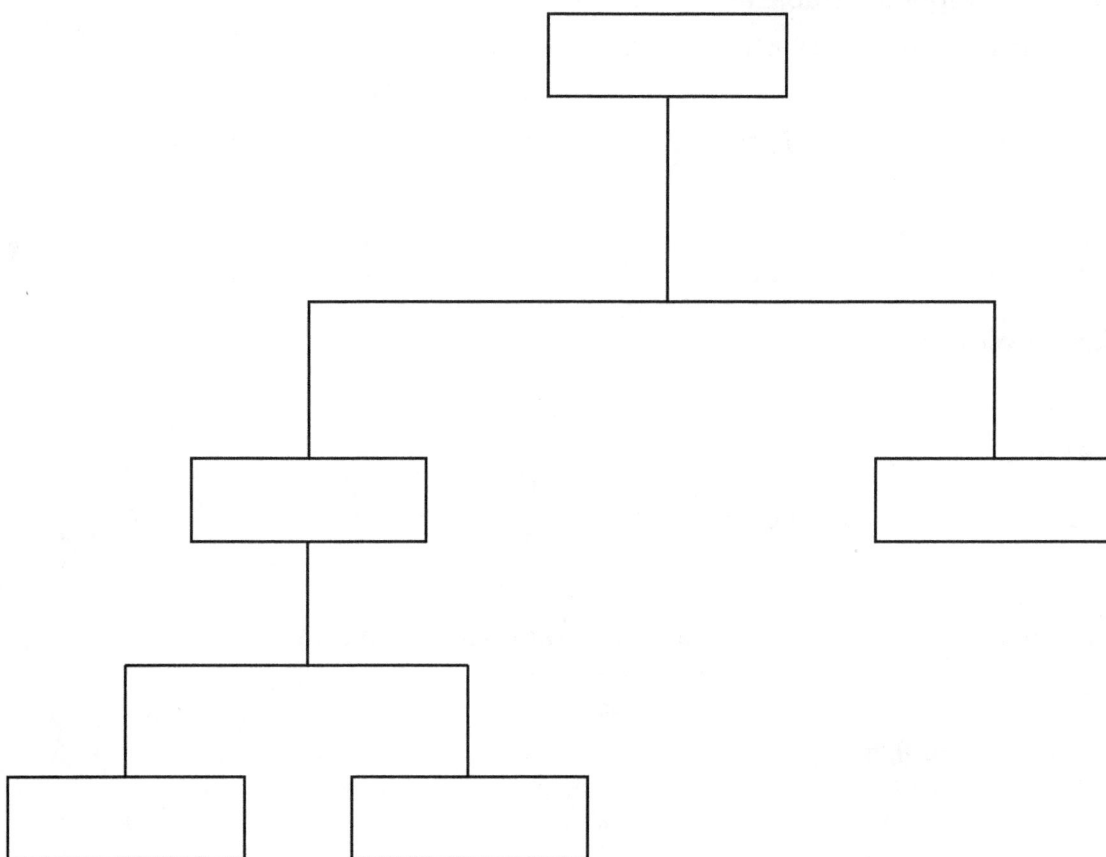

1.   **What's an electrolyte?**

2.   **What's a non electrolyte?**

3.   **What's a strong electrolyte?**

4.   **What's a weak electrolyte?**

ACTIVITY 5.0: Conductivity Test to Determine Whether Solution is an Electrolyte or Non electrolyte

Solution	Is substance an electrolyte or non electrolyte?	Is substance strong or weak electrolyte?
CH_3COOH (Vinegar) HCl (muriatic acid) NaCl (salt) $NaHCO_3$ (baking soda) CH_3CH_2OH (ethanol) $C_{12}H_{22}O_{11}$ (sucrose) NaOH (lye)		

POST-ACTIVITY QUESTIONS

1.   Why do some chemicals conduct electricity when dissolved in water?

2.   Why do some chemicals not conduct electricity when dissolved in water?

3.   Why do some chemicals when dissolved in water strongly conduct electricity?

How sodium hydroxide dissolves and dissociates into sodium and hydroxide ions

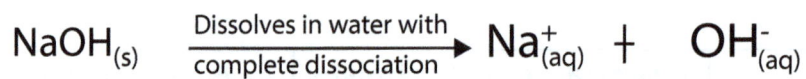

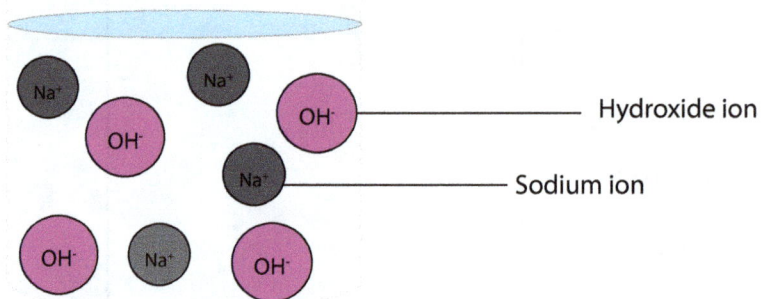

$$NaOH_{(s)} \xrightarrow[\text{complete dissociation}]{\text{Dissolves in water with}} Na^+_{(aq)} \quad + \quad OH^-_{(aq)}$$

Hydroxide ion

Sodium ion

How sugar dissolves and dissociates into neutral sugar molecules

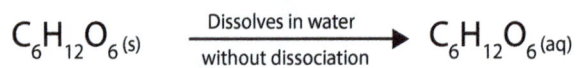

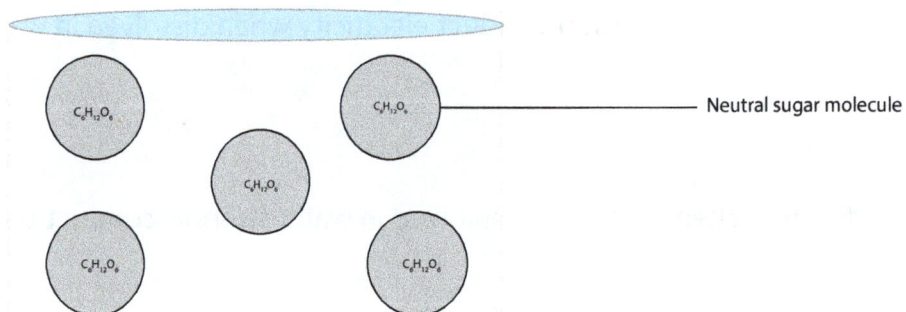

$$C_6H_{12}O_{6\,(s)} \xrightarrow[\text{without dissociation}]{\text{Dissolves in water}} C_6H_{12}O_{6\,(aq)}$$

Neutral sugar molecule

How hydrogen chloride dissolves and dissociates into chloride and hydrogen ions

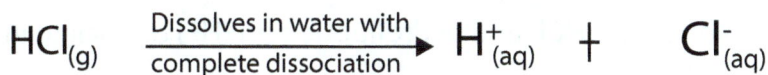

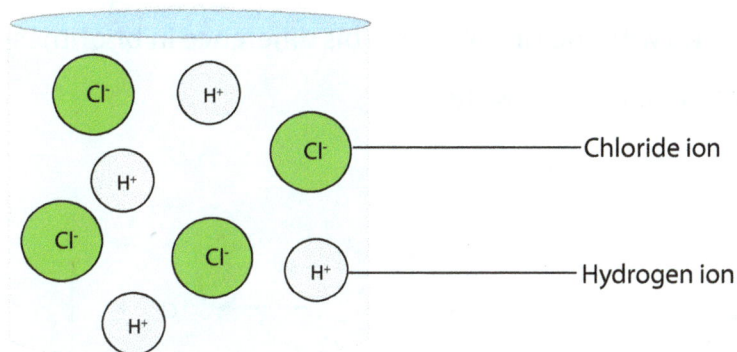

$$HCl_{(g)} \xrightarrow[\text{complete dissociation}]{\text{Dissolves in water with}} H^+_{(aq)} \quad + \quad Cl^-_{(aq)}$$

Chloride ion

Hydrogen ion

The H^+ ion produced from the dissociation of HCl has only one proton and no electron. Because of this, it's highly reactive. This high reactivity causes the hydrogen ion to immediately react with water to form hydronium ion. Here is an equation describing the dissociation of HCl in water: $HCl_{(g)} + H_2O ----> H_3O^+_{(aq)} + Cl^-_{(aq)}$

Here is Lewis dot diagram showing how the hydronium ion forms

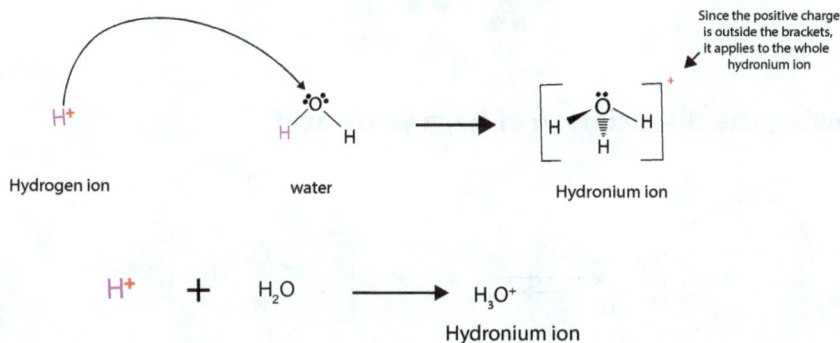

Since the positive charge is outside the brackets, it applies to the whole hydronium ion

Hydrogen ion   water   Hydronium ion

$$H^+ \quad + \quad H_2O \longrightarrow H_3O^+$$

Hydronium ion

From the diagram, the oxygen atom in H_2O shares its electrons with H^+ to form H_3O^+.

Why're some electrolytes strong and others weak?

A strong electrolyte dissociates completely in water, while a weak electrolyte partly dissociates. For instance, hydrochloric acid (HCl) and hydrofluoric acid (HF) dissociate in water to produce H_3O^+. However, if you test the conductivity of the same concentration of HCl and HF, you will notice that the light bulb of the conductivity tester glows brighter with the HCl solution than it glows with the HF solution. This difference in brightness is because of how well each chemical dissociates in water.

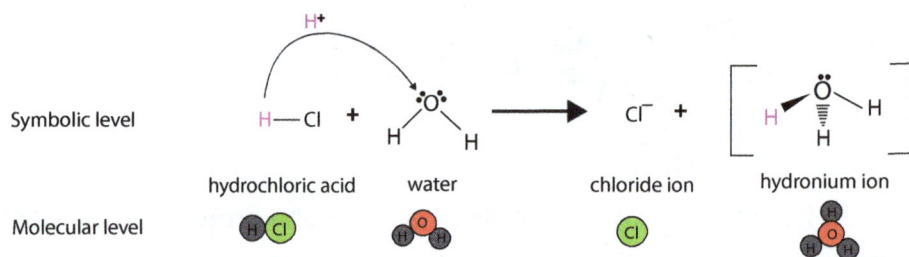

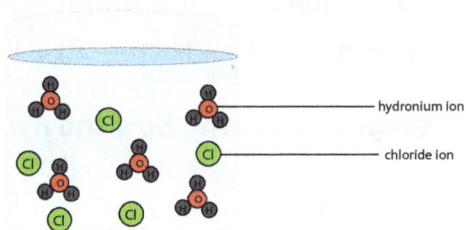

Here is a diagram showing the dissociation of HF, a weak acid

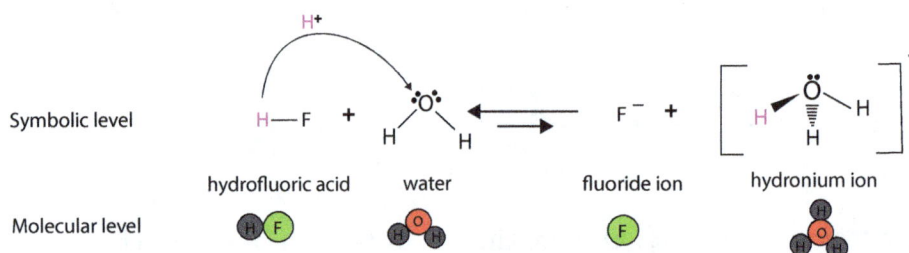

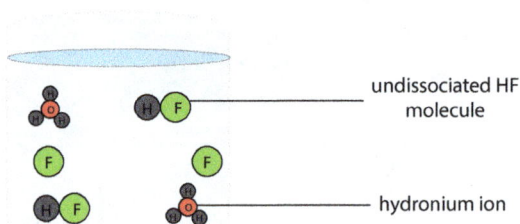

How does an acid or base behave at the molecular level?

What's Arrhenius definition of an acid?

An acid is an electrolyte that dissolves in water to produce hydronium ion (H_3O^+).

What's Arrhenius definition of a base?

A base is an electrolyte that dissolves in water to product OH^-, hydroxide ion.

What determines the strength of an acid?

The strength of an acid is determined by the extent of its dissociation in water. A strong acid dissociates completely in water, while a weak acid partly dissociates. For instance, hydrochloric acid (HCl) and hydrofluoric acid (HF) dissociate in water to produce H_3O^+. However, if you test the conductivity of the same concentration of HCl and HF, you will notice that the light bulb of the conductivity tester glows brighter with the HCl solution than it glows with the HF solution. This difference in brightness is because of how well each chemical dissociates in water. **Refer to the diagrams on page 106.**

Examples of strong acids

Formula	Name	hydronium ions produced by one molecule of acid
HCl	Hydrochloric acid	1
HBr	Hydrobromic acid	1
HI	Hydroiodic acid	1
HNO_3	Nitric acid	1
H_2SO_4	Sulfuric aicd	2

Examples of weak acids

Formula	Name	hydronium ions produced by one molecule of acid
HF	Hydrofluoric acid	1
HClO	Hypochorous acid	1
H_3PO_4	Phosphoric acid	3
H_2CO_3	Carbonic acid	2
CH_3COOH	Acetic acid	1

What's the difference between strong and weak base?

Similarly, the strength of a base is determined by the extent of its dissociation. A strong base dissociates completely in water, while a weak base partly dissociates. For instance, sodium hydroxide (NaOH) dissociates completely in water to generate sodium and hydroxide ions. While ammonia (NH_3) partly dissociates by picking up hydrogen ion (proton) from water to form ammonium ion (NH_4^+) and hydroxide ion (OH^-).

Here is a diagram showing the dissociation of NaOH, a strong base

$$NaOH_{(s)} \xrightarrow[\text{complete dissociation}]{\text{Dissolves in water with}} Na^+_{(aq)} + OH^-_{(aq)}$$

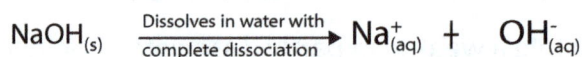

Hydroxide ion

Sodium ion

Here is a diagram showing the dissociation of NH_3, a weak base

Lewis dot diagrams showing hydrogen ion (proton) transfer from water to ammonia

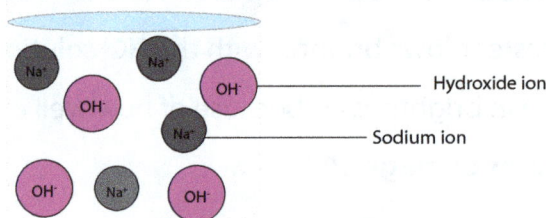

Left arrow is longer than the right arrow. This means the equilibrium lies more to the left.

water: proton donor

Ammonia: proton acceptor

hydroxide ion   Ammonium ion

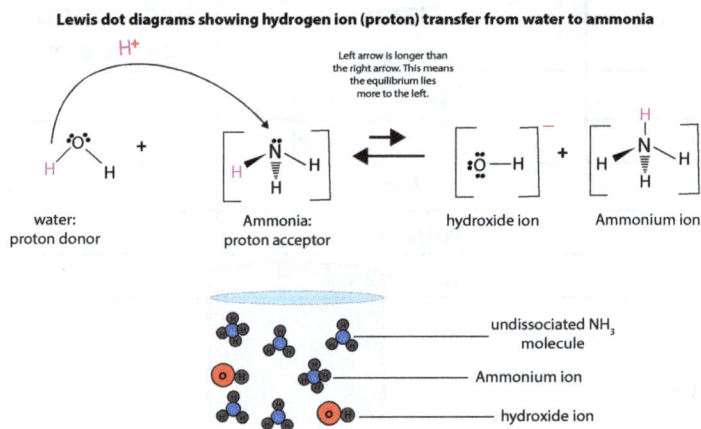

undissociated NH_3 molecule

Ammonium ion

hydroxide ion

Note!

Unlike sodium hydroxide, ammonia accepts a hydrogen ion (proton) from water to form ammonium and hydroxide ions. Because of this behavior, a base is generally defined as a substance that accepts a proton. While an acid is defined as a substance that donates a proton. These definitions are usually referred as **BrØnsted-Lowry** definitions.

Can water autodissociate? Yes, it does to form hydronium and hydroxide ions. Here is how:

Lewis dot diagrams show how one water molecule transfers its hydrogen ion (proton) to another water molecule

Long backward arrow means the OH⁻ and H_3O^+ ions easily combine to get back water molecules. The backward direction is favored and happens more often

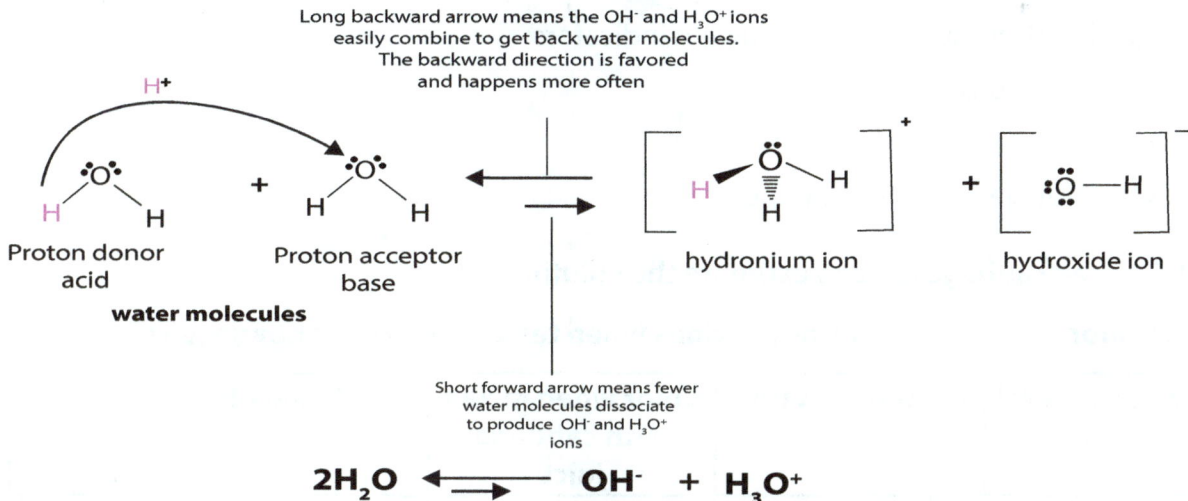

Proton donor
acid

Proton acceptor
base

water molecules

hydronium ion

hydroxide ion

Short forward arrow means fewer water molecules dissociate to produce OH⁻ and H_3O^+ ions

$$2H_2O \rightleftarrows OH^- + H_3O^+$$

What's the pH scale?

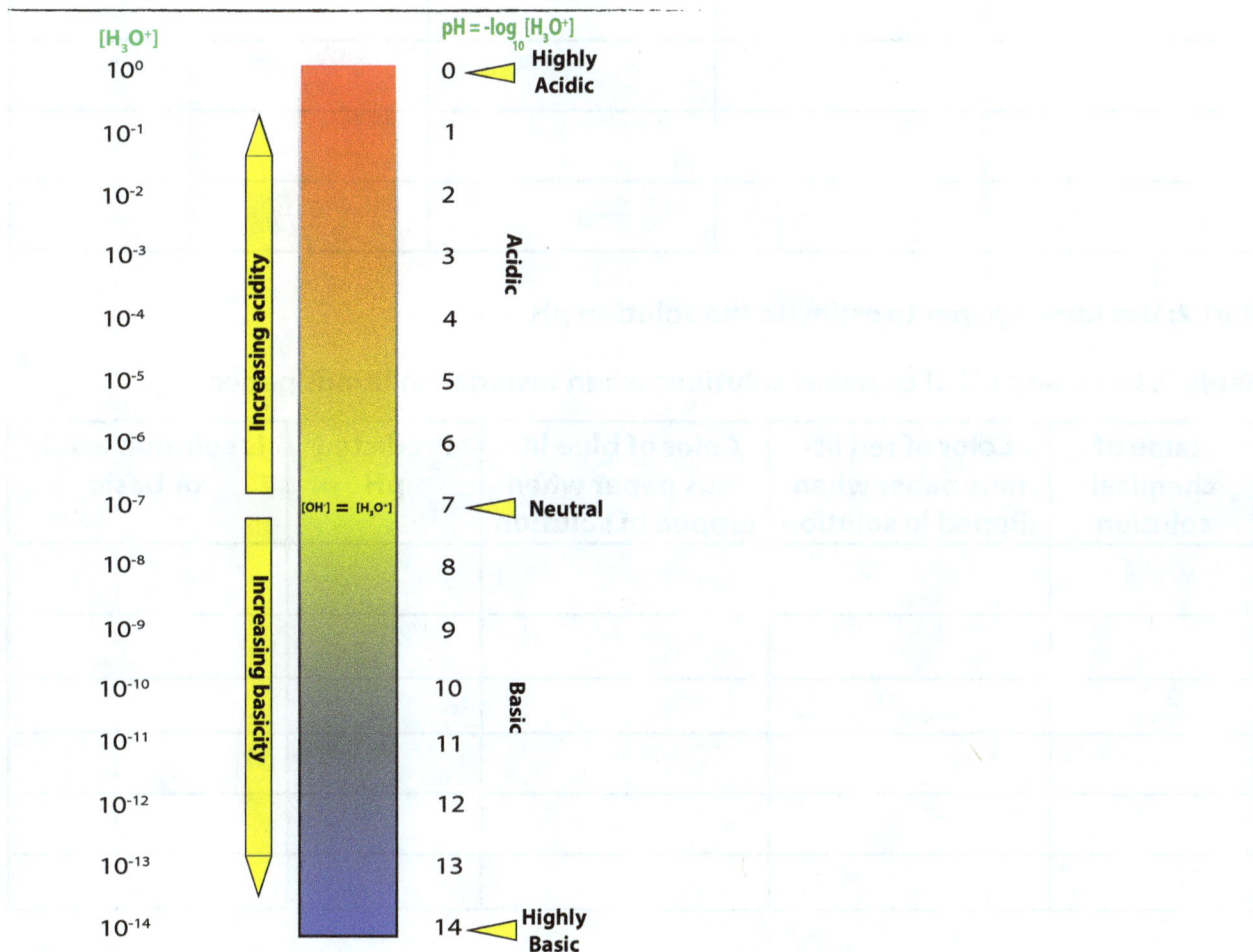

$[H_3O^+]$

$pH = -\log_{10} [H_3O^+]$

$[H_3O^+]$	pH	
10^0	0	Highly Acidic
10^{-1}	1	
10^{-2}	2	
10^{-3}	3	Acidic
10^{-4}	4	
10^{-5}	5	
10^{-6}	6	
10^{-7}	7	Neutral
10^{-8}	8	
10^{-9}	9	
10^{-10}	10	Basic
10^{-11}	11	
10^{-12}	12	
10^{-13}	13	
10^{-14}	14	Highly Basic

Increasing acidity

Increasing basicity

$[OH^-] = [H_3O^+]$

EXPERIMENT 6: DETERMINING THE pH OF COMMON HOUSEHOLD CHEMICALS

The pH of solutions can be measured in many ways. However, the accuracy of the pH depends highly on the method you use. In this experiment, you will determine the pH of common household chemicals by using the following indicators:

1.   Red cabbage juice
2.   Litmus paper
3.   Universal indicator paper

Part 1: Use red cabbage juice to estimate the solution pH

Table 1: Color and pH of chemical solutions when tested with red cabbage juice

Name of chemical solution	Color of solution	Color when mixed with cabbage juice	Predicted pH	Is solution acidic or basic

Part 2: Use litmus paper to estimate the solution pH

Table 2: Color and pH of chemical solutions when tested with litmus paper

Name of chemical solution	Color of red litmus paper when dipped in solution	Color of blue litmus paper when dipped in solution	Predicted pH	Is solution acidic or basic

Part 3: Use universal indicator paper to estimate the solution pH

Table 3: Color and pH of chemical solutions when tested with universal indicator paper

Name of chemical solution	Color of solution	Color of universal I. paper when dipped in solution	Predicted pH	Is solution acidic or basic

POST-LAB QUESTIONS

1.   From table 2, write an operational definition for an acid and a base.

2.   From your results, which method was best at measuring the pH of solution? Explain.

3.   Can you tell how acidic or basic a substance is just by looking at its color? Explain.

4.   From your results, classify the chemicals into three groups.

5.   Are all acidic and basic chemicals dangerous to your health? Explain.

GASES

LEARNING OBJECTIVES

After this lesson, you should be able to:

• describe the general properties of gases

General properties of gases

General properties of gases include the following. They can

• expand: gases expand to fill their container. Just imagine blowing air into a balloon. **Notice!** Gas molecules themselves do not expand. It's the distance between them and frequent collisions with the walls of the balloon that increase.

• take the shape of their container: Imagine blowing air into different shapes of balloon.

• be compressed: Just imagine the amount of air squeezed into a car tire.

• mix easily with other gases: just imagine the air around us. It consists of a mixture of gases: nitrogen, oxygen, carbon dioxide, and small amounts of other gases.

• exert pressure: feel your bike or car tire to see how hard it is.

Why do gases show these properties?

To understand why gases exhibit these properties, we must study gas behavior. To study gas behavior, we must consider four factors that can affect gas behavior. These factors are:

• **Pressure (P):** the force exerted by gas molecules as they strike a given surface.

• **Volume (V):** the space occupied by gas molecules.

• **Temperature (T):** A measure of how fast gas molecules move (kinetic energy).

• **Amount (n):** Amount in moles of gas molecules.

If any two of the four factors are related, a change in one produces a change in the other. To study the relationship between any two factors, we must hold the other two factors constant. For example, to study the effect of temperature on volume of gas, we must hold pressure and amount constant.

What's atmospheric pressure?

Atmospheric pressure is pressure exerted by air molecules as they **bounce off things**. Pay attention to text in bold "bounce off things."

How to measure atmospheric pressure

We can measure atmospheric pressure by a mercury barometer. A mercury barometer consists of a glass tube of at least 760 mm long with one end of the tube sealed and the other end opened. We can then fill this tube with mercury and invert the opened end into a beaker of mercury. This type of barometer, which was first designed by an Italian scientist called Evangelista Torricelli, is shown below.

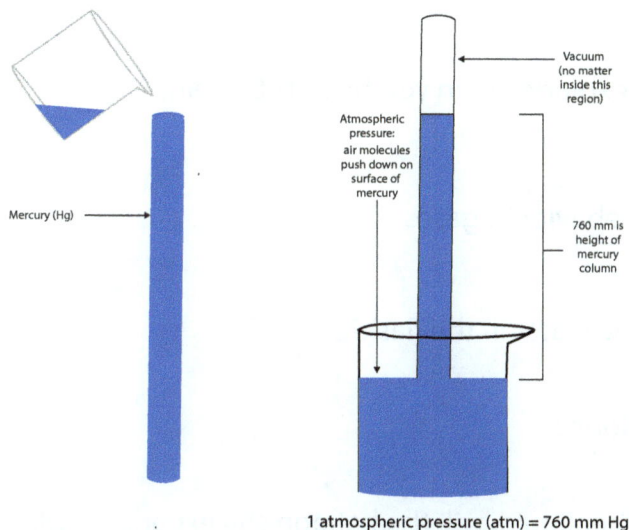

Vacuum (no matter inside this region)

Atmospheric pressure: air molecules push down on surface of mercury

760 mm is height of mercury column

Mercury (Hg)

1 atmospheric pressure (atm) = 760 mm Hg

As you can see, at sea level, the mercury level dropped to a height of 760 mm, creating a vacuum above the mercury in tube. Regardless of the diameter of tube, you will observe that mercury will drop to the same level: 760 mm.

Why did mercury drop to the same level?

The atmospheric pressure acting downwards on surface of mercury in beaker stops the mercury from draining out. This balance of forces is achieved when the **weight of mercury in tube is equal to the air pressure pushing downward on surface of mercury in beaker**. This pressure at sea level which supports a column of mercury at a height of 760 mm is called the **standard atmospheric pressure**. This standard pressure can be expressed in many units. These units include:

- mm of mercury (760 mm of mercury)
- atmosphere (atm), 1.00 atm
- torr (1 torr = 1 mm of mercury), 760 torr

From these units, we can write that : 1 atm = 760 mmHg

EXPERIMENT 7: EXPLORING THE PHYSICAL BEHAVIOR OF GASES

Caution: Wear safety goggles and be careful when placing the balloon over the mouth of the hot Erlenmeyer flask.

PREDICTIONS:

   a)   Will the balloon grow in size as the water in your flask boils? Explain.

   b)   Is energy added or removed from water molecules as the flask is heated? Explain.

   c)   Will a closed test tube explode when you heat it? Explain.

PROCEDURE

Develop a procedure to explore the behavior of gases

POST-LAB QUESTIONS

1.   Why was it necessary to boil the water in the flask?

2.   Why did the flask suck the balloon?

3.   If you replace water with another liquid, will the balloon still expand? Explain

4.   How will you remove the balloon inside the flask without using your hands to pull it off?

5.   What is the relationship between temperature and pressure of gas molecules?

6.   What is the relationship between temperature and kinetic energy of gas molecules?

7.   At constant atmospheric pressure, what is the relationship between temperature and volume of gas molecules?

8.   Do gas molecules have mass? Use evidence from this experiment to support your answer.

9.   How do gases exert pressure?

BLANK WORKSHEET

ENERGY

LEARNING OBJECTIVES

After this lesson, you should be able to:

- state and explain the two primary forms of energy
- state and explain the law of conservation of energy
- calculate the amount of energy a substance absorbs or transfers to another substance
- define and explain specific heat capacity

What's energy?

Energy is defined as the ability to do work. Energy is an abstract concept that is widely discussed in science. You will find it in chemistry, physics, biology and other areas of science. Because of this, it's sometimes called a crosscutting concept on which other concepts are developed.

What're the effects of energy on the properties of matter?

Properties of matter change when energy or temperature changes. These properties include pressure, density, electrical conductivity, hardness, and many more.

In fact, the flow and transformation of energy is what makes everything in the universe go. Without energy, the universe in itself dies.

What're the forms of energy?

Energy can exist in two main forms:

- Kinetic energy
- Potential energy

Kinetic energy is the energy an object has because of its motion, while potential energy is energy an object has because of its position relative to another or how its parts are arranged relative to one another. Of these two forms, energy in itself remains unchanged as it transforms from one form to another. And the total energy in a substance is usually equal to the sum of kinetic energy **plus** potential energy.

HOW DOES KINETIC ENERGY RELATE TO POTENTIAL ENERGY?

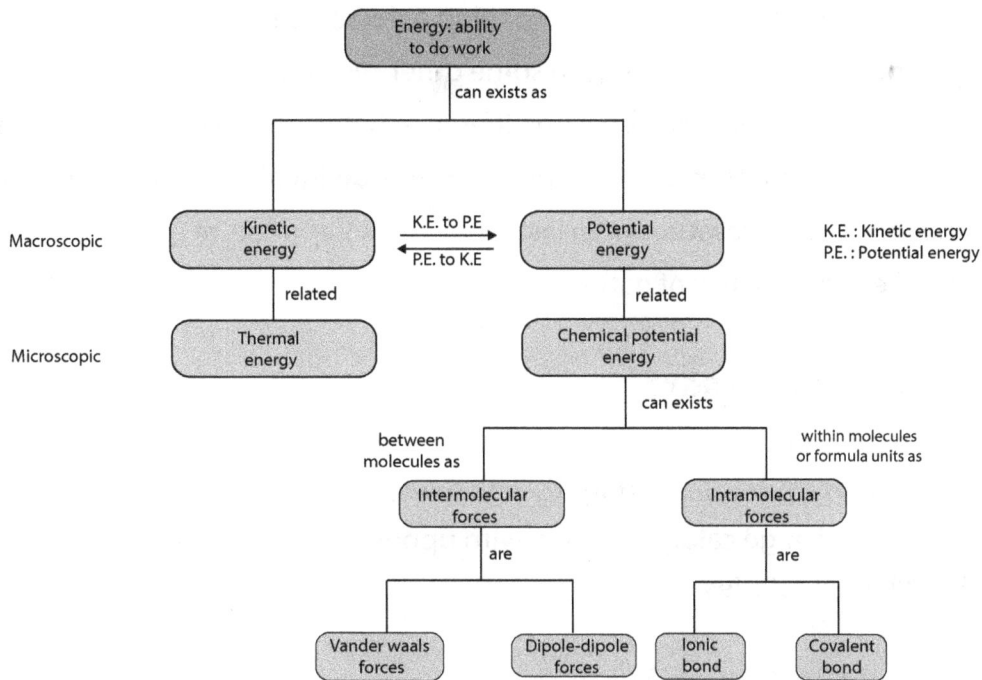

WHAT IS THE TOTAL ENERGY OF A SYSTEM?

It's always the sum of potential energy **plus** the kinetic energy

WHAT ARE THE SOURCES OF ENERGY?

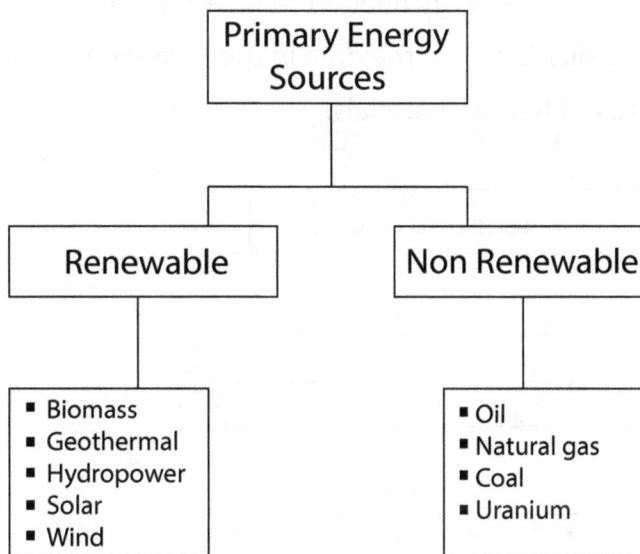

WHAT'S THE LAW OF CONSERVATION OF ENERGY?

The law of conservation of energy states that energy can't be destroyed or created, we can only transform it from one form to another. Meaning a decrease in energy in one system must be compensated by an increase in energy in some other system. If you reflect on the previous statement, you will realize that there is a core difference between the **law of conservation of energy** and **energy conservation**. **Energy conservation** involves the economic use of energy such that our energy resources can last longer for us and future generations. While law of conservation of energy is a law of nature.

WHAT ARE THE UNITS OF ENERGY?

We can measure energy in:
- calorie, cal (read small calorie, starting with lower case c)
- Calorie, Cal (read large calorie, starting with upper case C) or kilocalorie, kcal. Often called nutritional calories
- joules, J
- kilowatt-hour, kWh (units on our electric bills)

How are these units related to one another?

$$1 \text{ cal} = 4.18 \text{ J}$$

$$1 \text{ Cal} = 1000 \text{ cal} = 4180 \text{ J}$$

$$1 \text{ kWh} = 3.60 \times 106 \text{ J}$$

WHAT IS SPECIFIC HEAT CAPACITY?

Specific heat capacity is the amount of energy needed to raise the temperature of **one gram** of any substance one degree Celsius. Let's use the data in the following table to see how specific heat capacity vary across different materials.

Substance	Specific heat capacity $\left[\dfrac{J}{g \, °C}\right]$
Elements (metals)	
Silver, Ag	0.235
Copper, Cu	0.385
Aluminum, Al	0.903
Compounds	
Water, H_2O	4.18
Ethanol, CH_3CH_2OH	2.42

HOW CAN ENERGY BE TRANSFERRED?

Energy can be transferred by **heat** and **work**. Work is a way in which energy is transferred from one mechanical system to another. Heat is a way in which energy is transferred from a higher temperature to a lower temperature as a result of **temperature difference** between the **system** and **surroundings.**

There are three ways in which energy can be transferred by heat. These three ways are:

- **conduction, common in solids**

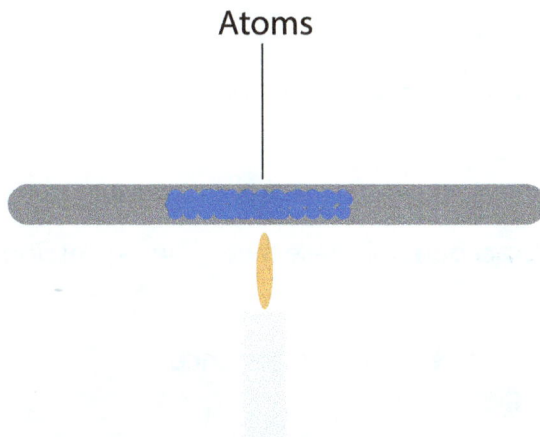

Atoms

- **convection, common in liquids and gases**

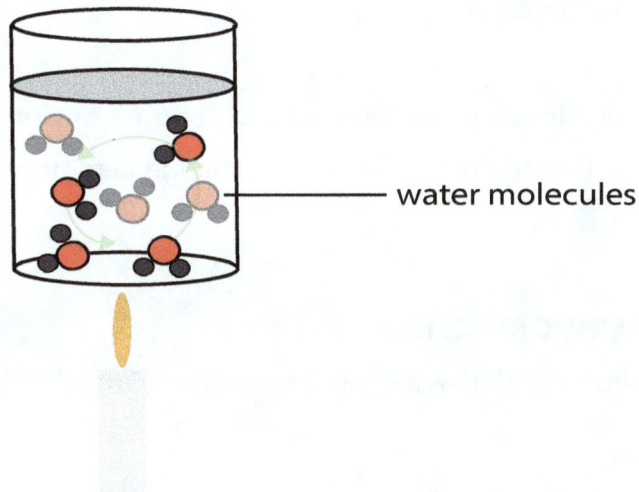

water molecules

- **Radiation**
This is how energy travels through vacuum. For example, energy from the sun reaches us by radiation

WHAT'RE SYSTEM AND SURROUNDINGS IN ENERGY TRANSFER?

To determine the amount of energy a substance absorbs or transfers to another, you must define your **system** and **surroundings.** The system is the substance you want to investigate and everything else that surrounds it is the surroundings. For example, imagine that you want to determine the amount of energy you will transfer to cold water in a bath as a result of temperature difference. In this case, you will be the system, and water, the surroundings.

How to calculate the amount of energy absorbed or released by a substance

From experience, we know that

- the longer we heat an object the hotter it becomes
- the more mass there is, the more energy we need to supply to heat the object

We also know that materials differ in how they absorb energy. Therefore, we can write a mathematical expression that takes into account all these facts. The expression is:

$q = m \times C_s \times \Delta T$

q = amount of energy absorbed or released by a substance

m = mass

C_s = specific heat capacity

ΔT = change in temperature

Note that to get change in temperature, you must subtract the initial temperature from the final temperature. Depending on the values, change in temperature can be positive or negative.

WHAT'S AN EXOTHERMIC REACTION?

An exothermic reaction transfers energy to the surroundings. An example is a burning candle.

WHAT'S AN ENDOTHERMIC REACTION?

An endothermic reaction absorbs energy from the surroundings. An example is a cold pack.

How to Calculate the Amount of Energy Absorbed or Released By a Substance

1.   The element aluminum has a specific heat capacity of 0.897 J/g °C. How much energy is absorbed by 30 g of aluminum if its temperature rises from 25 °C to 60 °C?

2.   Most cooking pans have a layer of copper on the bottom. If the specific heat capacity of copper is 0.385 J/g °C, how much energy is needed to raise the temperature of 135 g copper from 23 °C to 315 °C?

3.   If 50 g of hot tea cools from 80 °C to 22.4 °C, how much energy is transferred from the hot tea to the surroundings? (Assume that tea has the same specific heat capacity as water: 4.184 J/g °C).

4.   When one cup of popcorn burns in a bomb calorimeter, the temperature of water (1200 g) increases from 30 °C to 60 °C. Calculate the amount of food Calories in popcorn. Specific heat capacity of water = 1 cal/g°C.

5.   Design an experiment to determine the energy content in potato chips.

6.   Object A is hotter than object B, if the two objects are in contact, in which direction is energy transferred? Explain.

EXPERIMENT 8: HOW TO DETERMINE THE ENERGY CONTENT IN FOOD

How do food scientists calculate the energy content in food? In this experiment, you will determine the energy content in snacks.

PREDICTION: Rank the snacks from the highest energy content to the lowest energy content. Explain the reasoning behind your rankings.

PROCEDURE

Construct a ring stand to support a coke can containing 50g (50 mL) of water. Set your ring stand at a reasonable distance above the food sample so that the can will receive the most amount of energy as the food burns. Next, weigh your food sample and record the initial temperature of water. After recording the temperature, light the food and hold it underneath the can. Once the food stops burning, record the final temperature of water. For each trial, use the same amount of food and fresh sample of water.

Table 1: Energy content of food

Food Sample	Initial mass of food (g)	Final mass of food (g)	Mass of food burned (g)	Initial temperature of water (°C)	Final temperature of water (°C)	Temperature change of water (°C)	Energy content of food from experiment (Cal)	Energy content of food from food label(Cal)
							8	

POST-LAB QUESTIONS

1. Use the data from table 1 to calculate the energy absorbed by water in food **Calories**. This value is the energy content in snack. Write your answer in column **eight** of table 1.

2. Use the data on the snack label to calculate its energy content in Calories.

3. Use the results from question 1 to rank your food from the highest energy content to lowest energy content. Use the results from question 2 to rank your food from the highest energy content to lowest.

4. From your rankings, discuss the difference between the energy content calculated from the experiment and energy content from the snack label.

5. Since the system lost energy to the surroundings, what changes will you make to improve your results?

USE THIS WORKSHEET TO CONSTRUCT YOUR DATA TABLE AND DO YOUR CALCULATIONS

USE THE SAMPLE DATA TABLE TO LEARN HOW TO CALCUALTE AMOUNT OF ENERGY

Food Sample	Initial mass of food (g)	Final mass of food (g)	Mass of food burned (g)	Intial temperature of water (°C)	Final temperature of water (°C)	Temperature change of water (°C)	Energy content of food from experiment (Cal)	Energy content of food from food label (Cal)
Marsh.	6.2	5.4	0.8	24.0	31.0	7.0	0.350	2.76

Sample Calculations

Marshmallow: Initial mass (6.2g) – final mass (5.4g) = mass burned off (0.8g)

Final temperature (31°C) – Initial temperature (24°C) = temperature change (7°C)

To calculate the energy absorbed by water, you must multiply the mass of water by its specific heat and by its temperature change. Once you get your answer in small c calories, next, convert this value to big C calories. See the example below.

1. $Q = m \times c \times \Delta T = 50g \times 1.0 cal/g \times 7°C = 350 cal \times 1\ Cal/1000\ cal = .350\ Cal$

To calculate the label-derived energy content of each food sample, you must first get the Calories and the serving size in grams from the food package. Once you get that, divide the Calories value by the serving size in gram, and then multiply your result by the mass of food that burned off. See the example below.

2. $Cal/g = 100 Cal/29g = 3.45 Cal/g \times .8\ g = 2.76\ Cal$

Note that!

1000 calories = 1 Calorie

Specific heat of water (c) = 1.0 cal/g oC

NUCLEAR REACTIONS

LEARNING OBJECTIVES

After this lesson, you should be able to:

- define radioactive decay
- balance nuclear reactions
- describe the difference between chemical and nuclear reactions

Nuclear reactions are reactions that involve the nucleus of an atom. There are two types of nuclear reactions. The first is called radioactive decay and second nuclear transmutation. To understand these reactions, let's use the atomic model below to review the structure of an atom.

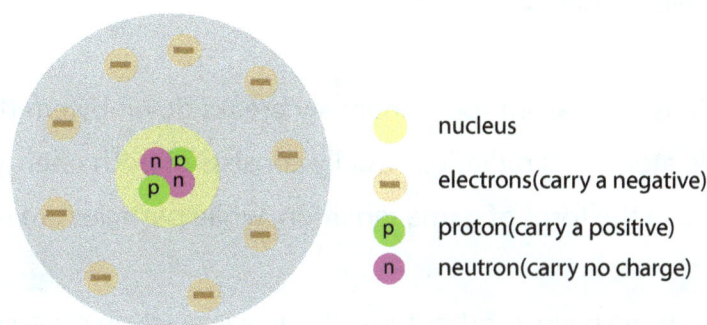

nucleus

electrons(carry a negative)

p   proton(carry a positive)

n   neutron(carry no charge)

From the model,

What are nucleons?

Nucleons are: protons and neutrons

What charges do nucleons carry?

Protons carry a positive charge

Neutrons carry no charge

In an atom, where are the nucleons?

The nucleons are in the atom's nucleus. The nucleus is surrounded by negatively charged electrons.

What is the charge of the nucleus?

The nucleus carries a positive charge. This is so because the nucleus consists of protons and neutrons. But the protons carry a positive charge, while the neutrons have no charge.

Is the mass of the atom concentrated in the nucleus? Yes.

The mass of the atom is concentrated in the nucleus. This is so because the volume of the nucleus is small and the combined masses of the proton and neutron is far greater than the mass of the electron.

What holds the nucleons together in the nucleus?

A force called a strong force holds them together in the nucleus

Can an atom become unstable?

Yes. If you consider an atom to consists of a system of moving charged particles, then you will realize that as an atom gets larger, any disturbance to these moving particles can cause an atom to become unstable. This instability can cause an atom to disintegrate so that its parts can rearrange to form a more stable atom. When this happens to an atom, we usually say that an atom has undergone radioactive decay.

What is radioactive decay?

It is a process where the nucleus of an atom spontaneously breaks down by emitting radiation in order to form a more stable atom. This radiation usually consists of both energy and particles. Energy is usually released in the form of gama radiation, while particles are released in the form of alpha and beta.

In addition to radioactive decay, we have another type of nuclear reaction called nuclear transmutation.

What is nuclear transmutation?

Nuclear transmutation is a process where an atom is bombarded with protons, neutrons or other particles to force it to disintegrate.

Properties of alpha, beta, and gama radiation

Properties	Alpha (He atom)	Beta (an electron)	Gama (energy)
Symbol	α	β	γ
Charge	+2	-1	None
Mass (amu)	4	0.0005	None
Ability to ionize	high	moderate	None
Ability to penetrate	low	moderate	high
Can be stopped by	paper	aluminum	lead

Here is a diagram summarizing radioactive decay

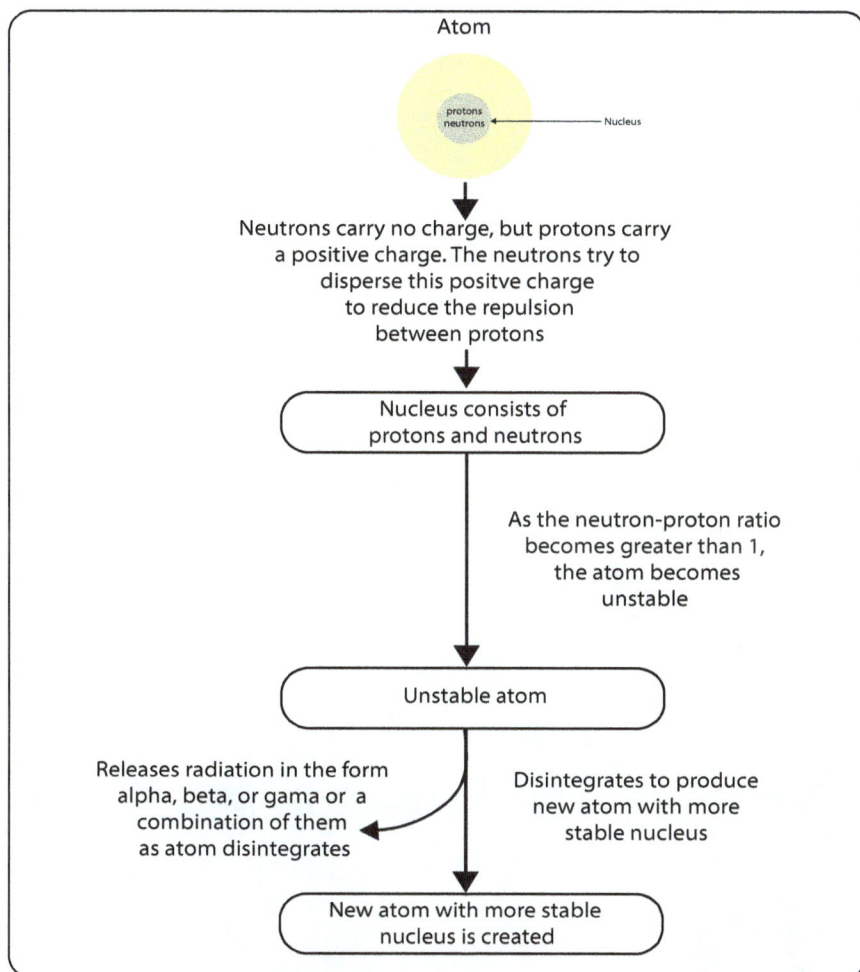

How to write and balance nuclear reactions

1. If radium-226 emits an alpha particle to form a new atom, write a complete nuclear equation for the reaction and solve to find the new atom.

2. Find the missing variable represented by the question mark in the following nuclear equations.

    a. $^{235}U + {}^{1}n \longrightarrow {}^{131}Sn + ? + 2\ (^{1}n) + Energy$. [Nuclear Fission]

    b. $^{3}H + {}^{2}H \longrightarrow {}^{4}He + ? + Energy$. [Nuclear Fusion]

Example 1

Example 2

QUESTIONS ON NUCLEAR CHEMISTRY

1.   Describe the effects of radiation on human health?

2.   Describe the applications of radioactivity?

3.   What is half-life?

4.   List at list three differences between chemical and nuclear reactions?

CHEMISTRY MISCONCEPTIONS RESEARCH PAPER

LEARNING OBJECTIVES

After writing your research paper, you should be able to identify:

- chemistry misconceptions common among students
- effective teaching strategies that can best address students' misconceptions

ASSIGNMENT

Identify any chemistry misconception appropriate to the grade level you wish to teach, and develop an effective instructional strategy to address this misconception.

DIRECTIONS

To start, first visit the Ohio Department of Education website at **http://education.ohio.gov/ Topics/Academic-Content-Standards/Science** to explore the state science standards and the misconceptions associated with the grade-level you wish to teach in future. Select one particular misconception you will like to address in your "future class" then use Google scholar (**http://scholar.google.com/**) to search for published work concerning that particular misconception. The published work may appear in peer-review journals like the ones I have listed below. You can find some of these journals by searching online through Dunbar library. If Dunbar library doesn't subscribe to the journal you want, you can always request an article through interlibrary loan, or you can simply ask your librarian at the Dunbar library. After reading your articles, answer the questions as outlined on the template. Use authentic sources only, these sources may include textbooks, websites, and journal articles. Cite your sources according to APA. This report is your homework assignment and will form 8% of your final grade. You must submit your final report in dropbox by the due date set by your instructor.

SUGGESTED JOURNALS TO HELP YOU GET STARTED

Science Education

Journal of Science Teacher Education

Journal of Elementary Science Education

School Science and Mathematics

Journal of Chemical Education

International Journal of Science Education

Journal of Research in Science Teaching

International Electronic Journal of Elementary Education

Chemistry Education Research and Practice

Journal of Science Teacher Education

The Science Teacher

SUGGESTED WEBSITES TO HELP YOU GET STARTED

http://education.ohio.gov/Topics/Academic-Content-Standards/Science

http://www.middleschoolchemistry.com/lessonplans/

http://www.acs.org/content/acs/en/education/resources/k-8.htm

FOLLOW THIS TEMPLATE TO ORGANIZE YOUR RESEARCH PAPER

(Include the questions in your paper)

1.   What are misconceptions?

2.   What are sources of misconceptions?

3.   How do students develop misconceptions in science?

4.   How do misconceptions affect learning?

5.   Which misconception are you trying to address?

6.   Why do students struggle to understand the chemistry concept behind the misconception you are trying to address?

7.   Describe in detail the teaching strategies you will use to address the misconception

   a.   Strategy 1

   b.   Strategy 2

8.   Why are some students' misconceptions difficult to dispel?

GROUP TEACHING DEMONSTRATION

LEARNING OBJECTIVE

After completing this assignment, you should be able to develop a chemistry lesson based on the 5Es inquiry learning model to address students' misconceptions in chemistry.

DIRECTIONS

Choose a chemistry concept and develop a 20 minute lesson plan appropriate to the grade level you wish to teach. During the last of week of class, you will teach your lesson such that students who hold misconceptions can develop a better understanding of the concept. Your lesson plan must include the common students' misconceptions you are trying to address. You can find chemistry lesson plans on the American Chemical Society website listed below.

EVALUATION

To assess your contribution to the project, you must evaluate one another on how well each member in your group contributes to the project. Also, during your teaching demonstration, your peers will evaluate you using the teaching demonstration rubric. Remember that this assignment forms 8 % of your final grade.

SUGGESTED WEBSITES TO HELP YOU GET STARTED

http://education.ohio.gov/Topics/Academic-Content-Standards/Science

http://www.middleschoolchemistry.com/lessonplans/

http://www.acs.org/content/acs/en/education/resources/k-8.htm

http://scholar.google.com/

PERIODIC TABLE

Key:

atomic number
Symbol
atomic mass

- Metal
- Nonmetal
- Metalloid

1 —— Group number

These rows go here →

1	2		3	4	5	6	7	8	9	10	11	12	13	14	15	16	17	18
1 H 1.01																		2 He 4.00
3 Li 6.94	4 Be 9.01												5 B 10.81	6 C 12.01	7 N 14.01	8 O 16.00	9 F 18.99	10 Ne 20.18
11 Na 22.99	12 Mg 24.31												13 Al 26.98	14 Si 28.09	15 P 30.97	16 S 32.06	17 Cl 35.45	18 Ar 39.95
19 K 39.10	20 Ca 40.08		21 Sc 44.96	22 Ti 47.88	23 V 50.94	24 Cr 52.00	25 Mn 54.94	26 Fe 55.85	27 Co 58.93	28 Ni 58.69	29 Cu 63.55	30 Zn 65.38	31 Ga 69.72	32 Ge 72.59	33 As 74.92	34 Se 78.96	35 Br 79.90	36 Kr 83.80
37 Rb 85.47	38 Sr 87.62		39 Y 88.91	40 Zr 91.22	41 Nb 92.91	42 Mo 95.94	43 Tc 98	44 Ru 101.07	45 Rh 102.91	46 Pd 106.42	47 Ag 107.88	48 Cd 112.41	49 In 114.82	50 Sn 118.69	51 Sb 121.75	52 Te 127.60	53 I 126.90	54 Xe 131.29
55 Cs 132.91	56 Ba 137.30	57-71 Lanthanoids	72 Hf 178.49	73 Ta 180.95	74 W 183.84	75 Re 186.21	76 Os 190.23	77 Ir 192.22	78 Pt 195.08	79 Au 196.97	80 Hg 200.59	81 Tl 204.38	82 Pb 207.2	83 Bi 208.98	84 Po	85 At	86 Rn
87 Fr	88 Ra	89-103 Actinoids	104 Rf	105 Db	106 Sg	107 Bh	108 Hs	109 Mt	110 Ds	111 Rg	112 Cn	113 Nh	114 Fl	115 Mc	116 Lv	117 Ts	118 Og	

57 La 138.91	58 Ce 140.12	59 Pr 140.91	60 Nd 144.24	61 Pm 145	62 Sm 150.36	63 Eu 151.96	64 Gd 157.25	65 Tb 158.93	66 Dy 162.50	67 Ho 164.93	68 Er 167.26	69 Tm 168.93	70 Yb 173.04	71 Lu 174.97
89 Ac	90 Th 232.04	91 Pa 231.04	92 U 238.02	93 Np	94 Pu	95 Am	96 Cm	97 Bk	98 Cf	99 Es	100 Fm	101 Md	102 No	103 Lr

134

www.ingramcontent.com/pod-product-compliance
Lightning Source LLC
Chambersburg PA
CBHW081544220326
41598CB00036B/6559